The Gardener's Guide to Growing TEMPERATE BAMBOOS

Michael Bell is President of the UK Bamboo Society, contributed to their booklet and regularly writes for the society journals. He holds one of the National Collections of *Phyllostachys* (cultivated forms) which forms part of his collection of over 165 species and forms and was the subject of a spectacular TV presentation. An engineer by training, Michael Bell has always appreciated the sophisticated structural qualities of bamboos and his interest in gardening with them has developed into a fascination for all sensual plants, particularly those that are found in the historic gardens of Cornwall, England.

Other titles in the Gardener's Guide *series:*

HARDY GERANIUMS *Trevor Bath & Joy Jones*
LILIES *Michael Jefferson-Brown & Harris Howland*
HELLEBORES *Graham Rice & Elizabeth Strangman*
FRITILLARIES *Kevin Pratt & Michael Jefferson-Brown*
IRISES *Geoff Stebbings*
CLEMATIS *Raymond J. Evison*
PENSTEMONS *David Way & Peter James*
HOSTAS *Diana Grenfell*
SALVIAS *John Sutton*
ASTERS *Paul Picton*
DAHLIAS *Gareth Rowlands*
DAYLILIES *Diana Grenfell*
PEONIES *Martin Page*
MAPLES *James G.S. Harris*
ORCHIDS *Brian & Wilma Rittershausen*
CANNAS *Ian Cooke*

The Gardener's Guide to Growing TEMPERATE BAMBOOS

Michael Bell

David & Charles

TIMBER PRESS
Portland, Oregon

PICTURE CREDITS

Neil Campbell-Sharp pages 72-73; John Glover pages 138, 146; Jerry Harpur page 77; Harry Smith page 19; plates by Karl Adamson, all other photographs by Marie O'Hara.

NOTE

Throughout the book the time of year is given as a season to make the reference applicable to readers all over the world. In the northern hemisphere the seasons may be translated into months as follows:

Early winter	December	*Early spring*	March	*Early summer*	June	*Early autumn*	September
Midwinter	January	*Mid-spring*	April	*Midsummer*	July	*Mid-autumn*	October
Late winter	February	*Late spring*	May	*Late summer*	August	*Late autumn*	November

Hardback edition first published 2000
Reprinted 2001
Paperback edition first published 2003

First published in the UK in 2000 by David & Charles Publishers,
Brunel House, Newton Abbot, Devon
ISBN 0 7153 0859 9 (hardback)
ISBN 0 7153 1581 1 (paperback)

First published in North America in 2000 by Timber Press Inc.,
133 SW Second Avenue, Suite 450, Portland, Oregon 97204, USA
ISBN 0-88192-445-8 (hardback)
ISBN 0-88192-570-5 (paperback)

A catalog record of this book is available from the Library of Congress

Designed and Edited by Jo Weeks
Illustrated by Coral Mula
Printed in Italy by Stige SpA

page 2 *Chimonobambusa tumidissinoda* (p.91) is the epitome of elegance in a light woodland setting.

page 3 *Pleioblastus shibuyanus* 'Tsuboi' (p.126) is one of the best variegated leaf bamboos.

CONTENTS

INTRODUCTION

The ageing wicket gate seemed to sense our hesitation. It resisted our efforts and groaned at the hinges, scraping on the compacted earth. It had taken us the best part of an hour to progress this far from the lane where we had first seen the 'Garden Open' sign. There had been the hazards of a field of cows to overcome, and the overgrown perimeter hedge with its many beckoning openings had laid many a false trail. The narrow path that we had at last found skirted the big house but did not seem to have the status of an entrance to a grand garden, even for us invading tourists, even for an estate a century past its prime and hidden behind a shroud of exuberent growth. Anticipation pulled us on.

Beyond the gate, the narrow, deeply sunken path curved, and was lost behind a huge beschorneria standing sentinel with its glaucous yucca-like leaves and its scarlet flower spike, arching threateningly overhead. The vertical line of the waist-high stone retaining wall was continued skyward on the other side by a breathtaking bamboo. It was not surprising that my wife, my two young daughters and I felt hesitant and intimidated under such majesty. Had I but known that this was to be not only the initiation to the rest of the garden, but also the foundation for a substantial portion of the rest of my life, then I would have stood awhile in awe. I gave the bamboo scant attention, however, as I hastened through the canyon.

When we found ourselves at the other side, we knew that this was going to be special. We had visited several other memorable gardens during our holiday, but we sensed an inherent, though elusive, quality here that raised this on to a higher plane. We followed the informal path through the cinnamon trunks of rhododendrons the size of trees. Branching low, their sinuous form layered naturally so that it was difficult to tell which was the original. I was firmly reminded of our place in the scale of time as we bowed low to follow the path under the dark canopy of branches. We emerged from submarine light into a sun-spangled glade between exotic trees. A group of tree ferns, their crown of huge feathery leaves way overhead, was enjoying the balmy summer air. Even to the disinterested these plants epitomize jungles and age and all things carboniferous. Age – and time, the fourth dimension – became one of the senses. Here was dappled sunshine, the rustle of leaves and the creaking of branches in the gentle breeze, perfume from flowers but mainly from the spice of resinous trees, the music of bird song, the tactile smooth bark and soft foliage. These bombarded our senses. Time, anticipation and imagination unified all to achieve that higher plane.

Then, around the bend, I saw the first of many bamboos that would reach my soul that day. In a garden of giants, suddenly I was alongside a plant such as I had never encountered before. Its height was meaningless as it vanished into the canopy above, and the diameter of its huge canes (which I later learned to call culms) was irrelevant. Primarily, here was a plant with quality. Every aspect of it stimulated the senses: it was sensual in form like a giant shuttlecock of sculptured quills; it was sensual to feel the smooth gloss surface and to hear the delicate rustle of leaves in the breeze like no other plant. The very thin-walled culms creaked with the slow movement and, somewhere above, gently

With its elegant form and bright colours, *Bambusa multiplex* 'Alphonse-Karr' makes a fine pot plant for a special spot.

knocked on tree branches or other culms like muted wind chimes. Beneath our feet was a carpet of rustling leaves reminiscent of the pleasures of autumn. But mostly it reminded us of far-off villages of huts, bananas and bamboos and other delights of the imagination. I later learned that this was *Himalayacalamus falconeri*, and it is still the largest example of that species that I have seen anywhere.

Then there was another plant, identical to the first. But no! As if to outdo it, this one had culms with random yellow, red and green stripes of an intensity that I had not seen on any other plant. Next I came upon a giant, with culms around which I could not close my fingers. By contrast it had delicate layered branches and tiny leaves. This was *Phyllostachys edulis* whose well-spaced culms invited us to walk between them. Past this was another bamboo of more compact form, and yet more beyond.

My daughters, who normally treated their father's recent interest in gardens with barely concealed disinterest, had gone – a sure indication that *this* garden's many charms were universally appreciated. They were in a world of make-believe, of jungles, dinosaurs, *Gulliver's Travels* and *Treasure Island*. A sunken road under a rustic bridge became the place of trolls; the aerial view from the bridge's span was the view from a mountain looking down on the lost world the other side.

Neither the garden nor I have aged for the better since that day, more years ago than I care to say. My daughters now have children of their own, and the *Himalayacalamus falconeri* flowered and died quite a few years ago, but I still make the pilgrimage to Penjerrick in Cornwall at least once a year to refresh my memories and to cleanse my soul.

When I returned to my small city garden, only a few hours' drive away, the austerity of its Japanese style seemed as sterile as a glass of water. How could I live with it after tasting wine? And how could I capture and reproduce all those memories in such a small area? There was only one plant that encapsulated most of the delights of that garden and that would thrive in my harsher conditions: this is how I came to plant my first bamboo. I convinced myself that, if I planted it in one corner, it would be completely compatible with the Japanese style, although in my heart I knew that my motives were totally alien to its theme. On that day I was seduced into the world of sensuous plants. I had bitten the apple, and the world would never be the same.

The bamboo grew apace; it dominated the garden. I acquired two additional different species, my interest growing as fast as my collection. I was fascinated with the bamboos' unique method of growth and their amazing impact, and, consequently, my Japanese style slowly died over the next few years.

When I probed the experts of the day for more information they inevitably retreated, until one suggested that I obtain a copy of a recently published book on bamboos by A. H. Lawson. The fact that my local library did not stock *Bamboos: a gardener's guide to their cultivation in temperate climates*, and that it took the bookshop five weeks to obtain a copy is an indication of the level of interest in those times. I was like a boy before Christmas as I waited for that day when it arrived and I collected it. But once back home, in a quiet room, I was transported into another world. Here were the bamboos that I had been growing, but now I had names. Here were those plants that I had first seen at Penjerrick, plus bamboos I had never even dreamed about. Above all, the book was packed with knowledge, first-hand experiences, and the information needed to recognize each plant. This book was to remain my bible for over 20 years while I enjoyed my hobby in an absolute vacuum, devoid of interest from outside. When at last I met others who shared my obsession, and all at once societies were formed, collections were restored and labelled, and good nurseries were established, I was amazed to find that, thanks to that book, every plant in my substantial collection had been correctly identified.

In his preface Lawson acknowledges his debt to that other great classic *The Bamboo Garden* by A. B. Freeman-Mitford, which was published in 1896, over 70 years earlier. Lawson explains that he had written '*Bamboos*' because there had been many introductions in those intervening years and much knowledge had been gained but not put into print. It is only just over 30 years since Lawson recorded his experiences but the same situation exists today as the subject has moved on considerably, particularly in the last 10 years. Lawson's book is still one of the best for the needs of the average

Phyllostachys edulis can eventually grow tall enough to tower over the surrounding trees, even in southern England.

gardener, but it is sadly out of print, very difficult to obtain and, naturally, in great need of revision.

Both authors made it clear that their work was not intended to be a botanic monograph, but, as Lawson wrote in his preface, 'primarily as a book founded on practical experience for those who are interested in growing these fascinating plants.' It is not my intention to elevate my work to the same status as these legends but this is exactly the niche that so desperately requires filling today, and I make no apologies for basing my writing upon their proven format. If this book is responsible for just one convert, who treasures it as much as I have my copies of '*Bamboos*' and *The Bamboo Garden*, then I will have been fully rewarded.

But a word of caution before we continue: bamboo nomenclature, and the frequent changes made to it, has been a minefield for the unwary and for those, like me, who are not trained in its complexities. Even those who are competent in the science have to remain alert to the constant alterations. However, as it is almost always only the generic name that changes, with a little flexibility it is easy for the horticulturist to rise above these problems. Fortunately the clouds are beginning to lift and some paths can just be discerned through the remaining wisps of mist, but it is inevitable that further changes in nomenclature will become necessary with time. I have made every effort to ensure that the names used in this book are in accordance with the latest thinking at the time of writing.

Pleioblastus hindsii is rarely grown to this impressive size, but it is a tough species that is suitable for most locations.

Bamboos are not like any other plant that I know. From their method of growth, through their visual contribution within the garden setting, to their flowering – almost every feature is unique. And they hold the key to many enigmas that are yet to be resolved. This alone would fuel the desire to know more about this group of plants, but in addition their charm and qualities are enthralling, so take care as you read on – these are sensual plants without equal and very liable to become addictive.

1

THE WORLD OF BAMBOOS

Bamboos are a sub-family of the extensive grasses family, Poaceae (Gramineae). The true grass family is probably our most important group of temperate plants providing us with the majority of our flora. It is also one of the largest groups of flowering plants, said to comprise over 600 genera and nearly 10,000 species. Although it cannot be called symbiosis in the strict sense of the word, the very close relationship between man and grasses must be one of the greatest influences upon our biosphere. Except for the very hostile regions of this planet, grasses grow and are exploited by us to such an extent that they form a very substantial part either of our diet or of the diet of the animal species upon which we depend. We encourage grasses to flourish by flooding fields, burning heaths and cutting down forests, so that they dominate vast areas and grow in much greater profusion than they would if it was left to nature. We select forms to meet all our needs: grasses provide most of our staple foods, including wheat, barley, oats, rice, corn, millet and even sugar, and they provide grazing and bedding for our animals. Nowadays we see huge plains of grazing animals or acres of cereal crops where once forests stood. In certain parts of the world terraced hillsides of paddy fields have become so familiar that they do not seem alien to the natural landscape and forests of sugar cane appear as intrinsic as reed beds.

Grasses and the Survival of Mankind

Grasses have characteristics that, whatever your religious persuasion, are difficult to describe without using the words 'God-given'. What would we do with the grain crops if they did not mature with the military precision that is within their nature? How could we make fields of grazing animals self-sustaining if grasses were not one of very few plants that relentlessly continue growing from the base, regardless of the amount of cutting at the top? If we did not have grasses, what other plant would sustain our requirements for large amounts of animal feed during hostile seasons without using additional precious land?

It is not just for food that we grow grasses. They also answer some of our practical needs. All over the world grass species of one type or another have long been used as roofing materials. Restios are used in the southern hemisphere, cultivated reeds in wetland areas and woody grasses on the open plains. There is always some grass species to serve us, whether hay to provide bedding, reeds (juncus) to make torches, miscanthus to weave baskets, or papyrus to construct rafts, and in our gardens grasses are cultivated for decoration.

Among the grasses, the bamboo family might be considered as the supreme provider. One plant could, if desired, provide forage for cattle, shoots to eat, all the materials needed to construct a house, utensils, armour and weapons. In fact bamboos could fill almost every need from cradle to grave, and indeed do in many parts of the world. After travelling across Asia in the 1920s, the American plant collector David Fairchild wrote, 'Take away bamboo from the Javanese culture and there would have been scarcely a house left standing, not many bridges, nothing to sit upon, nothing to carry water in, no hats, no fences, no erosion control, no birdcages, no scarecrows, no beds.' He continued with regret, 'Take away the bamboo from our civilization and we would simply have to go fishing with some other kind of rod.' And since then even this has been replaced by modern materials.

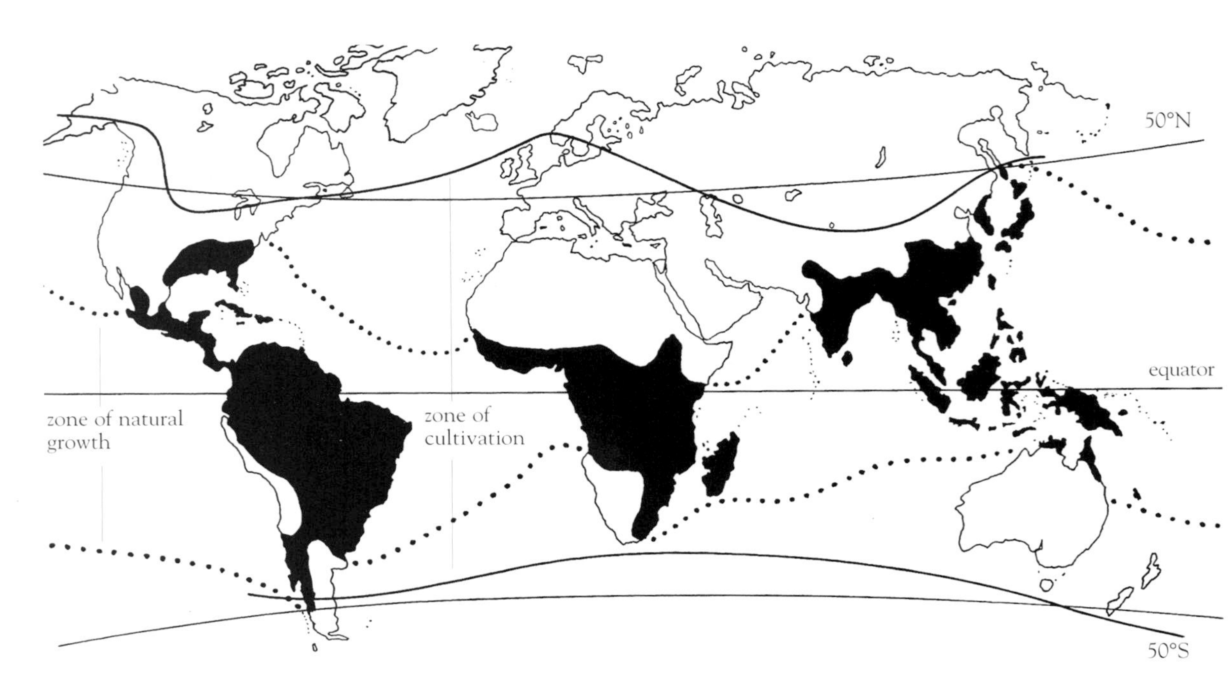

Map showing the natural distribution of bamboos, and their much wider spread in cultivation, which is particularly notable in the northern hemisphere.

THE BEGINNINGS

Grasses are ancient plants but, being soft, they leave few fossil records. We know from the tooth profile of extinct animal species that some were ruminants which ate grass, and it is generally agreed that herbaceous grasses or herbaceous bamboos in their present basic form evolved in the early cretaceous period (135–65 million years ago). This was the age of Gondwanaland, the supercontinent that disintegrated to produce all the major continents of today, which explains the almost universal spread of bamboos around the world in tropical and warm temperate regions. Bamboos must have already evolved from herbaceous grasses in the southern hemisphere before the splitting of the continents.

It is thought that the major bamboo lineages were already formed by the Miocene era (23.7–5.3 million years ago), and that the Tertiary period was one of great diversification within those groups. Fossil bamboos have been found recently in several countries, including Andance, south of Lyon in France, in Miocene strata dating from 6.5 million years ago, and also from the late Tertiary period (approximately 2 million years ago). At this time the continents had reached something like their present form, but the ice ages had not yet begun. DNA analysis reveals that *Arundinaria gigantea*, the only species growing in North America, is most closely related to *Pseudosasa amabilis* from China, half the world away. This distribution would seem unbelievable were it not also seen in genera such as magnolias and rhododendrons, and so remains indisputable evidence of a common origin.

We can surmise that the extent of the present range of bamboos has been curtailed by their inability to adapt to very arid regions or extreme cold. The ice ages, which affected the northern hemisphere, would have eliminated them from Europe and most of North America, and effectively isolated them within areas flanked by the natural barriers of the Himalayas and the arid regions of Eastern Europe. In North America *Arundinaria gigantea* remained and evolved several forms, and was until recently widespread in favourable regions.

While grasses adapted to colonize mainly the open habitats, the bamboos were plants of the forest. More recently woody bamboos adapted to semi-shaded or open habitats, although a few species and most of the herbaceous genera remain as forest floor plants. Where bamboos have also adapted to suit ecological niches varying from lowland warm temperate conditions to cold high-altitude zones, different species have given rise to many distinct forms.

DISTRIBUTION

Wild bamboos are very successful in all continents except Europe, but particularly so in the orient and South America. Until recently half the recorded genera and about half the known species were of Chinese origin. It has been discovered, however, that South and Central America have the greatest number of species and these are also the most diverse and interesting.

Taxonomists do not agree on the number of bamboo genera, but Ohrnburger and Goerrings in their extensive book *The Bamboos of the World* list 122, with 41 from the Americas, 16 from Africa, including Madagascar, 63 from southern and eastern Asia, and two from Australia. Undoubtedly there remain other genera to be discovered, and as large areas of bamboo country remain largely unexplored botanically, there are probably hundreds of species still unrecorded.

The natural range of the bamboo is enormous: chusqueas have been recorded growing almost up to the permanent snow line in the Andes, *Sasa kurilensis* can be found growing at a latitude of 46 degrees north, while *Chusquea culeou* can be found at 47 degrees south. The range in which they will grow is even wider: in the northern hemisphere they are now cultivated beyond 60 degrees north. In the southern hemisphere there are no large land masses further south than this natural range, apart from the rest of South America where bamboos are probably not cultivated.

MOUNTAIN SPECIES

Mountainous regions have an extremely diverse flora, a good proportion of which has adapted to temperate conditions. It is thought that flora in these areas evolved relatively recently, in parallel with the formation of these mountain zones. On most slopes of the Himalayas many more plant species, including a wealth of bamboos, can be found than there are native species in the whole of the British Isles. The cool, temperate parts of the Andes in Chile and Argentina have had very similar evolutionary pressures to the Himalayas and are another area of amazing biodiversity. They also possess a wide variety of bamboo species that probably have great potential in the temperate garden. There are only a handful of chusqueas in cultivation, but these include quite a few additional interesting natural forms.

MOUNTAIN BAMBOOS IN GARDENS

Generally most bamboos from mountainous regions are very well adapted to our conditions, and plants are available to suit the coldest of gardens, but bear in mind that success is not just a matter of temperature tolerance. Although they are cold, damp and misty, the homelands of these bamboos are often much closer to the tropics than are our temperate gardens, so sometimes these species require a higher light intensity than we can provide, and sometimes a cooler or more humid summer, too. Also it should be remembered that species from tropical latitudes which experience cold conditions usually do so on a daily basis, whereas those from colder latitudes do so on a seasonal basis. This means that they do not experience very cold soil temperatures or long periods of dormancy induced by months of cold. Some probably grow in permanently low soil temperatures from cold water run off from above – all conditions difficult to reproduce in some cultivation areas. There are those species that welcome free air movement and those that must have shelter. In their natural habitat, some species from the colder regions lose all their leaves during winter, and some have been reported as dying back to the roots each year, but this does not always occur in cultivation. Over-grazing is another problem often encountered in the wild but not in the garden, so with good care our plants can sometimes be more imposing than those growing naturally. The genera from these regions contain some of the hardiest bamboos in cultivation (-29°C/-20°F), ranging to many on the borderline of being classed as fully hardy.

LOWLAND SPECIES

Lowland China is the other major region where temperate bamboos are found, the main genus being the extensive and very distinctive *Phyllostachys*. Here, winters are cold and summers are hot, and as a consequence members of this group are also some of the hardiest that we grow (-26°C/-15°F). Most phyllostachys grow smaller and more compact in regions with cooler summers, but this is no great disadvantage in the average garden. It is usually possible to find species that will make an impressive grove even in a more northerly garden, and the genus is mostly very tolerant of greater heat and dryer air than the mountain species.

Oriental island species

The islands of Japan and Taiwan contain many bamboos that are very hardy and, although these regions mostly have hot, humid summers, the plants from these regions are very adaptable. The genus *Sasa* is to be found extensively throughout Japan, even as far north as the inhospitable Kuril Isles. It covers large areas, defeating all competition. It brings this property to our gardens, together with its tropical appearance. Most should be introduced with great caution for they are frequently seen dominating the vegetation. In the wild it often becomes brown and desiccated during the winter, but as a garden plant it generally remains much more attractive.

Tropical species

Other than the South African *Thamnocalamus tessellatus* all the species from Africa, Madagascar, Australia, southern and eastern Asia, including the Pacific Islands, and most from South and Central America are tropical. They contain many very interesting and varied plants but are not within the scope of this book as they cannot be grown successfully in a temperate climate.

EVOLUTION

Palms and cycads were contemporaries of the first grasses and are part of our flora to this day. Much like the crocodile in the animal world, they have not changed substantially over time. Grasses and bamboos, however, have evolved and adapted to survive, and are considered to be among the more scientifically advanced of plants. Their small flowers have almost dispensed with most of the unproductive features designed to attract pollinating insects, and have clearly evolved from more elaborate structures. Petals are reduced to two small scales, needing a hand lens to observe properly, a far cry from their forebears. The running rhizomes of some temperate species are also considered to have developed from tropical clumping species and it gives them a considerable advantage over their predecessors. Other features are not so clear cut, however as primitive characteristics, such as having few branches, are often combined with the more advanced running rhizomes.

The very long periods between flowering of most species, and the very poor seed production of others, can be considered evidence that bamboos no longer have a great need to produce a new generation from seed with all its inherent risks – seeds and seedlings are very vulnerable to the elements and to predators. Poor flowering and seed production would wipe out less successful plants. *Bambusa vulgaris* has now been introduced into most of the warm regions of the world and is one of the most common species to be found in the tropics. In spite of, or maybe because of, all this attention from humankind, only the occasional flower has been observed, and no seeds have ever been recorded. Its spread has relied entirely on vegetative propagation. The lack of flowers is probably because it has evolved into a very successful species and needs to go through the rather vulnerable cycle of reproduction only under extreme circumstances. Perhaps repeated vegetable propagation by mankind and the selection of non-flowering or readily rooted clones is a factor in some geographic regions, but in others it has been allowed to run wild with none of these influences.

ADAPTABILITY

Bamboos are very adaptable plants, which contributes greatly to their success and could also be considered a feature of advanced evolution. Most plants occupy a niche in the environment where they are more successful than other species. Outside this niche they quickly show signs of stress and are overtaken by species more suited to the conditions. Some are vigorous and cover large areas, while others are specialized and local, but few are able to adapt their form to suit the prevailing conditions. A fern can be seen on a sunny wall or in a shady wood but it is fundamentally the same, the only variation being its degree of luxuriance. Bamboos, by comparison, alter their form not only to accommodate extreme conditions but to thrive in them. By means of their tenacity and adaptability, many species have evolved from life in the shade of

An upright clone of the very variable, but always impressive, species *Chusquea culeou*.

Fargesia nitida is deservedly a most popular species, combining elegance with a tough constitution and great hardiness.

woodland into thriving on the edge of woodlands or in glades. It then required only minor evolutionary changes for them to become plants of the open countryside, provided there was protection from the wind, and plenty of water.

Low light levels

While exploring the extremities of a neglected but once glorious old garden, I came across a conifer forest that had probably been planted about fifty years previously. It was silent and dark within, and the only plants to be found in these testing conditions were cushions of moss and a few diminutive ferns, all other plants having expired long ago as the heavy canopy of leaves closed overhead. To my great surprise, growing in these conditions, so inhospitable that every other cultivated plant had died, including almost all of the native species, were the bamboo remnants from what must have once been part of the grand garden. The first plant I discovered was *Fargesia nitida*. It was not the small, poor, starving specimen that you would expect, but one of the finest and most stunning bamboos that I have ever seen. Its normally small leaves had been reduced still further to about half the size, its culms had increased in diameter and height, and their purple colour was masked by a glorious blue-grey bloom. The most suprising thing, however, was that the normally compact rootstock was open. It had grown to cover a large area and comprised hundreds of plants, some with room to walk between. This was not the *Fargesia nitida* that I knew, and at first I thought that it was a new and very distinctive clone. I expended much energy transplanting a good piece into my garden only to find, to my disappointment, that within 12 months it was just like all the other plants of this species.

Fargesia nitida is a shade-loving plant of the forest, so perhaps it could be expected to adapt to these extreme conditions, but there were other bamboos less obviously suited to the site but still thriving. *Phyllostachys* is a genus that usually requires strong light but here plants were growing in quantity. Although they were only a few metres tall, they were dark green and luxurious. *Pleioblastus simonii*, another plant recommended for open conditions, was also growing smaller than normal, but very green and very elegant in contrast to its normal rigid utilitarian style.

There were five species in all, each adapting in its own way and colonizing a large area. Not only had they endured where all other plants had failed, but they had changed their form and prospered. Many species of bamboo will grow in woodland conditions, waiting for trees to fall in order to achieve their full potential, but these plants were so healthy that they had little need for the canopy above them to open. They had little need, either, to produce seed that could lie dormant in anticipation of such an event like other plants might. They had evolved on a short-term basis, not just to survive, but to prosper, in conditions where newly planted specimens would quickly fail. When one takes into account such variation in form due to growing conditions, and consider, too, the various clones of some species that are in cultivation, it is not suprising that incorrect identification of bamboos is common: great care is needed when putting a name to an unlabelled specimen growing in a garden.

Their ability to adapt may be one reason why bamboos have such long intervals between flowering. I know of a specimen of *Phyllostachys aurea* that, when found in dense woodland, was over 100 years old but was only 2m (6ft) tall and had only about six culms. Within two years of thinning the trees, this same specimen was a vigorous plant three times the height and comprising dozens of culms, having had a head start on all its competitors when conditions improved. It would have been a disaster for it to have flowered during the dormant phase before the trees were cleared. If it had been strong enough to form seeds, which is most unlikely, they would not have germinated, and the effort of flowering almost always kills any weakened specimens such as this was.

Bogs and plant competitors

One of the finest plants of *Phyllostachys nigra* 'Boryana' that I have seen had adapted to growing in a bog when a leat contouring the hill above burst its banks and the water seepage remained unnoticed for many years. Conversely, the largest *Chusquea culeou* at Pitt White in Uplyme, England, which was tended by A. H. Lawson, is growing in a very dry position in shallow soil over stone. Although these are not conditions to be recommended, when a bamboo is growing strongly it has the ability to prosper in adversity.

I have observed *Sasa palmata* dominating those thugs of the plant world *Rhododendron ponticum* and laurels. Normally about 2m (6ft), *Sasa palmata* can, if it finds dense shade, grow to twice that height. When it finds good growing conditions, it puts out leaves to form a complete canopy over the competition below, be it laurels or seedling trees. In conjunction with a remorseless rhizome system and closely packed culms, this canopy can outrun any opposition and kill a forest in one tree generation. A plant like this is not one to accept that it

Sasa palmata 'Nebulosa' can quickly create a tropical atmosphere in the coldest regions, but is extremely invasive.

is beaten and it will also produce seeds so that it can colonize a more favourable location. Remember this, and treat most of the sasas with great respect if you are bold enough to introduce them to your garden. This type of rhizome is not called 'guerrilla' without good reason. When it is growing strongly I know of no method of permanently removing *Sasa palmata* from a garden setting without total destruction. At the other extreme, I have seen it no more than 50cm (20in) high, but still very healthy, on windswept moorland.

Identification problems

From this it can be seen that it is difficult to distinguish clones from such temporary variations, and growing conditions need to be considered when identifying plants 'in the field'. Variations are often unique to a species, for instance, in an open site *Chusquea culeou* can be observed to have short branches on the windward side and longer branches on the lee. I have not seen this in other species. Species with running rhizomes are often compact in cool or less-than-ideal growing conditions, and top growth can vary between rigid and lax in many species. It is important to remember that these different growth patterns are not as a result of damage, as with most plants growing in inhospitable conditions such as the gnarled growth of trees in windswept locations. Instead, the modified growth pattern of bamboos seems to be an automatic response to enable it to prosper.

To add to these problems of identification, many species come from steep mountain valleys where isolation has evolved many different forms and sub-species. There are many different forms of *Thamnocalamus spathiflorus* in cultivation and almost every plant of *Chusquea culeou* has very different features. Often seedlings are also very variable, much more so than those of most other plant species, so propagators carefully inspect every seedling and capitalize on any variation, as has happened with *Fargesia murieliae* recently where a multitude of named clones were marketed. For these reasons, precise identification should use botanic features, not physical features.

FLOWERING

All grasses have evolved a very precise flowering cycle. Leading expert on grasses Roger Grounds states that 'The flowering of grasses is ... something of a curious phenomenon, few other plants being quite so rigidly set in their habits.' Not only is each species selective to the days of the year, but also to the hour of the day, being influenced only slightly by external factors, such as adverse weather. Many species flower in the morning, others in the afternoon, and some twice a day, but in each case the period is surprisingly short. Roger Grounds suggests that the reason may simply be that, as wind pollinators, they have evolved to flower at the times most propitious for pollination but concluded that '... it may be infinitely more complex. The reasons for the rhythms of flowering in grasses are not yet well understood.'

Although it was not his intention, Roger Grounds' comments could well be expanded to embrace the flowering of bamboos, with the timescale adapted to suit long-lived perennials. A wind-pollinated plant flowering even slightly out of sequence to its near neighbours would be restricted to self-pollination, a pointless process, or be eliminated in one generation. Considered in this light, the enigma of bamboos flowering at long intervals seems to make more sense, even if the process is not fully understood: they flower infrequently because they find little need to produce frequent seed and they flower in unison because, without a unifying force, the longer the period between flowering, the less likely it would be that a near neighbour would be in flower at the same time. It is also less likely that hordes of insects or animals would be ready to devour a once-off bonanza.

Grasses are considered difficult to cross-pollinate, but this is probably more to do with different flowering cycles and the very short pollination period than any incompatibility. It was recorded some time ago by the respected American scientist Floyd McClure that bamboos can cross with sugar cane, and more recently there have been reports of cross-pollination with rice – two very different species. This raises questions about the authenticity of some bamboo plants. For example, it changes the bi-generic cross x *Hibanobambusa* (p.105) from a seemingly unusual occurrence to a likely possibility, which would be limited only by the infrequent flowering of the parents, and gives weight to the theories of some experts that the genus *Semiarundinaria* is a cross between *Phyllostachys* and *Pleioblastus* (it certainly shows the characteristics of both and, although *Semiarundinaria* plants have flowered on numerous

Elegant *Fargesia murieliae* is ideal in this Japanese garden. It will remain reliably compact without constant attention.

occasions, I know of no records of seeds being produced, although this is not exceptional in bamboos). It also raises interest in making further crosses between bamboo genera in the enclosed world of the garden. More fundamentally, it raises questions about whether bamboos that have never been recorded as flowering can truly be called species (a species surely being a plant that reproduces itself consistently from one generation to the next in the wild). Some consider cross-pollination to be an important evolutionary process; if this is so it makes these chance occurrences significant.

STRUCTURE

The structure of bamboo, which is a 'tree grass', is remarkably sophisticated. Comparing it with a tree is like comparing the minimalism of a yacht with the mass of a galleon. A tree stands solid against the forces of nature. To support its veneer of living material, it needs a great mass of accumulated dead wood, regardless of its potential for decay and, hence, its premature demise. A bamboo moves with the wind, dissipating its forces, particularly those of sudden gusts that would otherwise cause stresses several times greater than those under steady conditions. With today's technology how best would you design such a structure? The tapering tube of a modern yacht mast is perfect, but nature did this millions of years ago with the bamboo cane, and also thickened the cane wall towards the base, where the stresses are greatest. A mast can buckle under extreme conditions. To prevent this, if we could, we would sub-divide it with reinforcing diaphragms. Again, nature did this millions of years ago with the bamboo cane, and it even varied the spacing of the diaphragms to suit the stress graduation. Metal alloys used in masts are homogenous and can, therefore, snap when they buckle. Carbon fibre is one of the latest, and best, materials for preventing this. The high-tensile longitudinal fibres of a bamboo culm are set in a softer matrix and are structurally very similar to carbon fibre. On a strength to weight basis, bamboo canes are stronger than steel, and this is just the supporting structure, which is comparable with the heart wood of a tree. However, unlike a tree the supporting structure of a bamboo is also living material. Its surface is very hard, so it is also the equivalent of the cambium and its bark, which in a tree offer no support. In addition the bamboo culm contains chlorophyll, as do the leaves.

Climbing up a screen *Chusquea valdiviensis* gives light shade. The culms rising overhead are, infact, branches.

Each culm does not need to stand alone in resisting stress like a tree does. A tree is damaged if it comes into contact with its neighbour. Its brittle branches easily break, letting in decay. The branches of a bamboo are made of the same resilient material as the culms, with a proportionally smaller void, and are very strong. Species with few large branches intermesh with others and with the surrounding tree canopy. Other species rest against each other on a cushion of leaves and branches. By these means, when the stress of a gale reaches the base, it has been dissipated and the mat of interlocking rhizomes and shallow roots have no need to produce structural taproots merely to support huge masses of rigid branches above.

All structural components of bamboos are of a segmented, modular design and if a part suffers damage it dies back only as far as the next module. Sometimes culms break under the loads of the wind or, more usually, the weight of snow, and when this happens, the sections do not part and die. Longitudinal splits are produced along the fibres so that transpiration continues through the damaged culms, until replacements grow and take over. With the help of all of today's computers and technology, it would be difficult to conceive a more poetic design, and like all structures born of exacting design principles, bamboos have an inherent beauty, instantly felt by all.

DIVERSITY

The same exacting design principles mentioned above reduce any variations between species to a minimum as changes to the ideal are unproductive, unless they accommodate different environmental pressures. There are some very interesting and quite startling variations, however, mainly found in tropical species that have evolved to suit particular situations.

In cultivation, and only just hardy enough to be classed as temperate, is *Chusquea valdiviensis* from Chile. When seed was collected and sent to Europe, it came with an alarming warning – the species was found dominating all other vegetation and producing vast impenetrable thickets 'like a giant bramble'. It is a representative of the many climbing bamboos found particularly in South America, and has amazing energy. Initially it produces relatively small, backward inclined branches at each node, which act like grappling irons on the surrounding vegetation. Soon, however, a huge secondary branch, at least as big as the main culm, is produced from within each grappling iron. This behaves just as the rooted culms, producing initially small branches, then a huge branch of its own, and so on. It can root at the tips of the very fine culm ends, which hang down like vines, and it is probably just as well that this species is usually severely checked by winter conditions even in mild temperate areas.

In cultivation in southern parts of Ireland, and perhaps elsewhere, is *Ampelocalamus scandens*, which is even less hardy. In its native China it produces very long culms of small diameter, that hang down moist cliff faces in a most elegant way, but its growth is much more conventional in the harsher conditions in which it is normally grown in cultivation.

STRANGE BAMBOOS

Surprisingly, there are many herbaceous bamboos. They are generally small, elegant and sometimes delicate species from tropical woodland regions of high humidity, and most normally flower and set seed almost every year. How they became classified as bamboos is unclear to me as they seem to break almost every layman's definition of a bamboo. They are outside the remit of this book.

Probably the most bizarre woody species is *Glaziophyton mirabile*, which was initially identified as a rush. It grows 2m (6ft) high in stiff reed-like clumps. Only a few of its culms ever have leaves, and then just one or two. It has no nodes other than one or two closely packed at the extremities, which are easily overlooked, and the main part of the stem is reinforced by inert bands of pith. For 15 years, the sterile culms of this strange plant were observed in inaccessible mountain tops in Brazil until eventually they were cut to the ground by fire, when normal-branched, bamboo-type culms appeared together with huge flowers, identifying it as a species of bamboo.

Most members of *Neurolepis* look similar to large clumps of grass. A species from Columbia has very long leaves with no branches, just like the new culms of the giant reed, *Arundo donax*, and *N. aperta* has no visible cane, looking superfically like an agave or an amaryllis. In some species the leaves approach 4.5m (15ft) long. *Rhipidocladum* has thin-walled culms, with up to 2m (6ft) between nodes, and *Arthrostylidium schomburghii* is reported as going even further with internodes up to 5m (16ft) long. One *Puelia* species produces root tubers, while another has a single huge leaf on top of a long slender culm. *Melocanna baccifera* produces fruit like a small pear which starts shooting even before the seed drops to the ground. Species of the herbaceous *Pharus* and other genera have flower spikelets with hooked hairs to enable them to cling to passing animals for easy dispersion.

VARIEGATION AND VARIATION

Variegations in the pigment of the culms of bamboos does not always seem to affect adversely the vigour of the plant, or its ability to survive in the wild, presumably because its main chlorophyll source is in the leaves. Reports of large areas of the brilliantly variegated *Bambusa vulgaris* 'Vittata' and *Guadua angustifolia* 'Striata' growing in the wild are too numerous to be explained as intervention by mankind. Harder still to explain, and certainly not influenced by cultivation, are reports of a wild *Chusquea scandens* with variegated leaves in an inaccessible region of Columbia. The Japanese book *Nikon Chiku-Fu* (written by Katayama Nawohito and translated as *The Cultivation of Bamboos in Japan* by Sir Ernest Satow in 1899) records naturally occurring variegated bamboos, such as the chien-tao-chu (striped bamboo) which grows in the mountains of Two Cheh.

Strange forms are also recorded by the same writer although these are not known in the West. A form called so-shi-chiku has flattened culms with branches on both sides, and bifurcates near the top. Some variation in the branching pattern occurs over the height and it would seem that these are, in fact, two culms fused in some way. Another form has culms that are forked, sometimes at the lower nodes and sometimes higher up. Known as futamata-dake, it can have two or more leaders. A whole plantation of the double-forked form was recorded as growing in the grounds of a monastry at Hangchow. A photograph in the magazine (Vol. 15, 1, 1999) of the Pacific Northwest chapter of the American Bamboo Society seemed to indicate that this is an abnormal branching arrangement.

Among the other interesting bamboos mentioned in *Nikon Chiku-Fu* are a sakasa-dake or upside-down bamboo that is artificially produced and difficult to visualize. There are also tales of a huge bamboo found at Lo-fu that was 6m (20ft) in girth, with 39 nodes, each 6m (20ft) in length, and another with a diameter of over 2.2m (7ft) and leaves as large as a banana. One of these would certainly steal the show at your local garden club!

USES

Mention bamboos to most people and they might think of arrows, or blow pipes, or even of oriental horror stories of men being tied down over fast-growing shoots. They may remind gardeners of the plant hunter George Forrest's escape from hostile Tibetans in the early 1900s when a plant hunting trip went wrong. His foot was impaled by a spike of bamboo and he hobbled for weeks before reaching safety. We may also think of the demise of the pandas and the habitat destruction that isolates them in small vulnerable pockets of bamboo monosystems. There are many other bamboo-dependent ecosystems and animals worldwide in a similar situation usually caused by forest clearance.

PLATE I

Bamboo culm colours

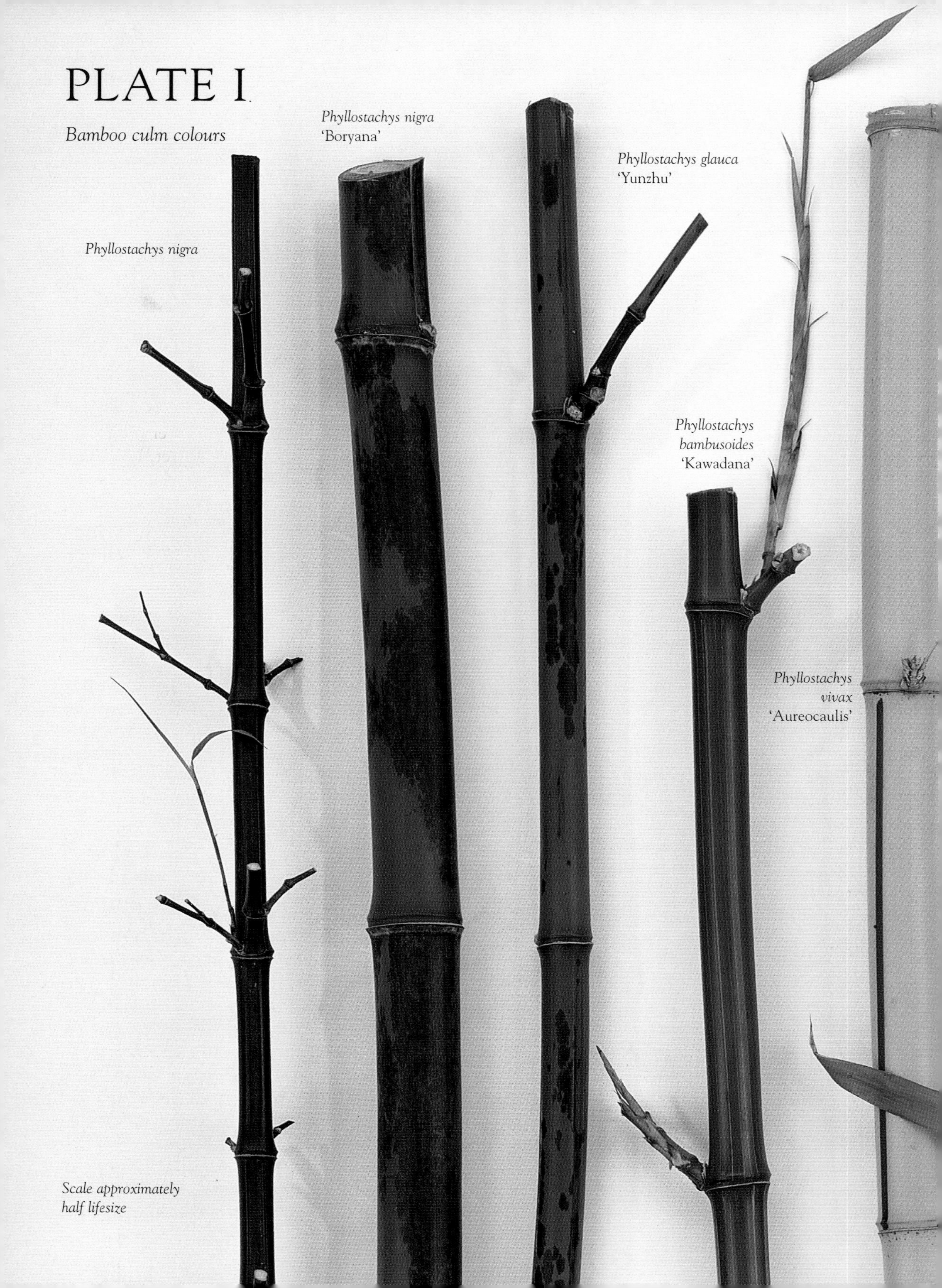

Scale approximately half lifesize

Phyllostachys aurea 'Koi'
Semiarundinaria yashadake 'Kimmei'
Phyllostachys bambusoides 'Castillonis'
Phyllostachys sulphurea 'Sulphurea'
Himalayacalamus falconeri 'Damarapa'
Phyllostachys aureosulcata 'Spectabilis'

Bamboos are also associated with extraordinary events in world history. It was in a staff made of bamboo that silk worm eggs were first smuggled out of China. The staff belonged to two monks and they took the eggs to the court of Emperor Justinian of Byzantia in AD552. The monks could not have foreseen the impact that their actions would have – that they were laying the foundations of the silk industry for the whole of the Western world.

On a more mundane level the majority of everyday items in the Far East are crafted from bamboo, just as timber was used in other regions before the industrial revolution. Strong canes are used as structural components for bridges, house building, and so on, and the leafy branches are used for thatching. The Chinese have used the stems as spears and to make guns. Sections of large culms are used as buckets and cooking utensils, and smaller ones as cups or as disposable containers for cooking small items in the embers of the fire. It is well documented that, since at least 200BC, the Chinese have drilled for oil to depths of 1000m (3500ft) using bamboo canes.

The variegated, dwarf species *Pleioblastus variegatus* benefits from pruning to keep the height in proportion to its spread.

Pipes, music and furniture

The techniques for processing bamboo have been improving since the dawn of civilization. To mention only one is unjust but the most ingenious is usually also the simplest. The techniques used to manufacture pipes are many and varied, as pipes are a basic need for all rural communities. Bamboos make ideal if short-lived pipes so long as the dividing diaphragms at each node can be removed. To break them from each end would limit the length of the pipes considerably, therefore the diaphragm is usually removed by cutting a local hole above each node and treating the culms as troughs. However, in one part of the world the diaphragm problem has been ovecome in a neat, if bizarre, way. A rodent is trapped in one end of the culm. Being unable to get a bite on the curving inside wall, it is forced to eat through each diaphragm to make its escape at the far end some while afterwards.

Among the more artistic uses for the bamboo are musical instruments, such as mellow Andean pipes and Japanese string instruments, fashioned from bamboo. Even more exotic are the instruments made by the natives of Malacca in Malaysia. They know how to pierce the culms of living bamboos in such a way that they become living flutes, capable of producing varying notes as the wind blows through the clumps.

All westerners are familiar with bamboo furniture (not to be confused with ratan furniture made using a species of climbing palm found in many tropical countries). Slated blinds, room dividers, screens and fences are all commonly imported. And we use bamboo canes in our gardens and as walking sticks.

Food and medicine

An elixir obtained from the deposits found within the culms of many bamboos has been prescribed in China and India for thousands of years as a cure for many ills. This was ridiculed by science until recently, when the elixir was found to have properties of absorbing internal poisons and acting as a catalyst.

Bamboo shoots are harvested before they break the soil surface and form a substantial part of the diet in Asia. The outer sheaths of the shoot need to be removed and most species require pre-boiling to remove the bitter taste. All species of bamboo require boiling with several changes of water to remove cyanogens.

The wine bamboo *Oxytenanthera braunii* is cultivated by a tribe in Tanzania for its alcoholic properties. A proportion of the new shoots have the top 6cm (2½in) cut off when they reach 60cm (2ft) high. On each of the following six days a few millimetres are removed from the cut top. The stump is then left for two days, before a container is lashed to it to collect the sap that has accumulated. Between a quarter and a half a litre is collected in the morning and, by removing a further few millimetres from the stump at midday, a similar amount is collected by the evening. It seems that the women and children drink the freshly tapped wine while the men let it brew for 12 hours to produce a strong wine. One family is capable of harvesting several hundred culm stumps a year. That is about 500 litres (100 gallons) of wine.

In gardens

Beyond being used for garden decoration, the practical employment of bamboos in the cooler, temperate regions is considerably more limited. Commercial use is not viable mainly because of the slower growth of the plants and the relatively high cost of labour. Plants are considerably smaller than, for example, the giant *Dendrocalamus giganteus* of Burma, at 35m (120ft) high and 30cm (12in) in diameter. The majority of cultivated temperate bamboos are remarkably uniform in style and features, which adds a restful quality to our gardens. Different clones, species, or even genera can, with care, be grouped together in a setting that imbues the garden with tranquillity and harmony. Most bamboos have similar requirements, but they do originate from a wide range of growing conditions around the globe, and if we are to understand our plants and make the most of them we need to study them in greater detail.

BAMBOO GARDENING – A HISTORY

Gardening started in China around about 2000BC, so, although there are no written records, it is certain that bamboos were in cultivation hundreds of years before they were grown in the West.

Chinese gardens started as idealized visions of paradise, drawing upon the rich variety of native plants, but later evolved into a more flamboyant style. After the visit of their first Envoy to China in AD607, the Japanese drew heavily upon this Chinese style, refining it to reflect their love of nature. The range of plants used by the Japanese, however, was narrow and conventionalized, and the use of bamboos was very limited. Forms were selected and nurtured as curiosities but had little use in the garden. Despite this, many species were introduced from China into Japan and became naturalized; indeed the most common bamboo in Japan today is *Phyllostachys edulis*, a Chinese introduction.

The naming of plants at this time was largely based on physical appearances: similar-looking plants were classed as regional variations of a species; our system, devised by Linnaeus in the mid-eighteeth century, now classifies these as different species. Natural variations were given completely different names so that, over the passage of time, their origins were lost and because of the absence of flowers, which would enable us to identify them, were irretrievable.

Nineteenth-century Europe

China had traded in small quantities with the West for hundreds of years but had largely been self-sufficient, having little to learn from the West and little need for Western goods. The Chinese were at their zenith during the eighteenth century and were hostile to foreigners, making exploration almost impossible. Plant collection was limited to those specimens that could be obtained from around the ports, and, therefore, plants that were already in cultivation in China. By the early nineteenth century, corruption and inefficiency had set in, with opium being traded by the West to pay for an insatiable demand for all things Chinese. By the time China decided to stop this illegal trade, Western technology had overtaken that of China, and Britain had supremacy of the seas. The ensuing wars ended in 1842 with the West winning concessions that effectively opened up the Chinese interior to exploration, and freed trade. From 1842 onwards British, French and American collectors invaded the Himalayas and the trickle of imported plants became a flood. However, due to their large bulk and infrequent and unreliable seeds, bamboos were not the ideal plants for collectors, so they did not feature greatly in the finds of the famous explorers.

Pleioblastus pygmaeus (old form) was popular in shady rockeries, such as this secluded dell at Heligan in Cornwall.

The zeal of the early Spanish and Portuguese missionaries had upset Imperial Japan. As a result, Japan was also closed to the West from 1612, except for a tiny trading enclave allowed to the Dutch. In 1853 all restrictions were hastily removed when an American fleet entered Tokyo Bay to negotiate trading agreements. It is logical, therefore, that the first introduction to Europe was *Phyllostachys nigra* in 1827, and sometime later *Arundinaria hookeriana* (now *Himalayacalamus falconeri* 'Damarapa'), as although the former is classified as the type plant of a species, both are in fact cultivated forms. Species were not introduced until the mid-1840s when *Himalayacalymus falconeri*, *Phyllostachys viridi-glaucescens and Indocalamus tessellatus* were sent to Europe.

The tropical *Bambusa vulgaris* had been spreading worldwide long before this, because of its practical uses and ease of propagation, and it had already reached Florida in the early nineteenth century. However, the main period of new introductions was the 1860s and 1870s with many new species reaching France, Belgium and England. M. Latour-Marliac of Lot-et-Garonne, France, was, at this time and at least up to the end of the century, classed as the greatest European importer of bamboos. From the records of the period, he was the major exporter to the rest of Europe, and also sold widely to other parts of the world. In France Eugene Mazel was constructing his now world-famous garden at Prafrance.

This was the period when, within Europe and America, the craft of gardening, as distinct from the art of gardening, had reached its zenith. The great wealth created by the industrial revolution was funding ceaseless and bewildering new imports of novelties, and vast sums were spent on elaborate bedding schemes and prestigious greenhouses. Interest in new exotics, such as bamboos, was tempered by the huge volume of other new material: records reveal that it was not until the 1890s that there was general interest in bamboos.

The first overall botanical treatment of the bamboos was published by horticulturist Colonel Munro in

1866. Auguste Rivière was director of the Trial Garden in Hamme, Algeria, and then head of the Luxembourg Gardens in Paris. He had grown and studied bamboos all his life, but died prematurely, just as he was about to publish his life's work on the subject. Fortunately, his son Charles completed and published *Les Bamboo* in 1890. This book thoroughly explored the anatomy, classification and propagation of bamboos and was unequalled for many years.

THE 1890s

Bamboo gardeners of Europe in the late nineteenth century were faced with many dilemmas. These new exotic-looking species were of unknown hardiness and were initially only planted in the milder regions of Europe, where they remain to this day in greatest profusion, for no good reason. Even *Sasa palmata*, the toughest of them all, was first planted in the Temperate House at Kew. It is now very widespread in cultivation, and is completely out of control in many gardens.

Plants came with various regional names, written in various characters. Labels were lost, those that were unreadable or not understood were thrown away, and it would not be cynical to suspect that fictitious names were added on occasion in these years of great competition. In *Nikon Chiku-Fu (The Cultivation of Bamboos in Japan* p.21) the author says about Western gardeners in 1899 that, '... it is not to be wondered at that gardeners and cultivators should find it difficult to determine the plants which are sent to them from this country [Japan]. They arrive usually in poor condition and three, perhaps four years may elapse before they develop sufficiently to allow of their being recognised. In the meantime, however, they have been named by the dealer, sometimes in a manner that leads to confusion. Often labels become illegible in transit, or being detached by accident, are afterwards assigned to the wrong plants.'

So confusion reigned from the start, probably not helped by the huge numbers of imports that arrived almost continously. The garden diaries from the great estates of the day reveal that at least one batch of bamboos was received during most years over the last decade of the nineteenth century. Caerhays Castle in Cornwall, for instance, is typical with records of five species added during 1895, eight during 1896, three during 1897 and one in 1898. Some recorded names are unknown today, and many specimens were unknowingly duplicated under different names.

During this closing decade of the nineteenth century, various other activities indicate a great and widespread enthusiasm for bamboos. The garden at Prafrance was bought by the botanist Gaston Negre whose family still own it. Kew Gardens concentrated their collection in its present location. Messrs Rivière finally published their famous *Les Bamboo*. W. J. Bean composed a comprehensive work in *The Gardener's Chronicle*. A. B. Freeman-Mitford published his book *The Bamboo Garden*, and Sir Ernest Satow published his translation, already mentioned, of a Japanese work on the subject, *The Cultivation of Bamboos in Japan*.

But even as this enthusiasm was spreading, undercurrents of change were taking place with the writings of gardeners such as the Irish William Robinson and the English Gertrude Jekyll. Under their influence there was a growing rejection of the values of these exotic gardens, and a return to the simpler cottage-garden style. Coinciding with a decline in the wealth of the great estates at the turn of the century, a mood for change swept the Western world. Gradually records of anything relating to bamboos ceased. In England this happened almost overnight, but interest lingered on for a few years in other parts of Europe with, among others, Jean Houzeau de Lehaie from Belgium. Houzeau was very influential and produced many important publications between 1872 and 1916.

NORTH AMERICA

In North America there has always been more emphasis on the commercial production of bamboos for timber, paper-making or shoots. This is mainly due to the fact that in some areas the climate is quite suitable for producing large culms. The first recorded importations of bamboos were *Pseudosasa japonica* in about 1860, *Pleioblastus simonii* in 1876 and *Phyllostachys aurea* in 1882. During the same period Henry Nehrling, Curator of Milwaukee Museum of Natural History, bought a tract of land near Orlando and here large shipments of bamboo, both temperate and tropical, were planted. The garden and house remain to this day.

The great American plant collector David Fairchild made extensive contacts in Japan during the early years of the twentieth century. With his partner, Barbour Lathrop, and his prodigy, Frank Meyer from the

Even in cooler areas, the vigorous *Phyllostachys nigra* 'Boryana' can form groves if it has good growing conditions.

Netherlands, he contributed a great deal to what was known about bamboos in that period, establishing a collection near Brooksville, Florida, and several introduction stations. He was told of a large bamboo grove near Savannah, Georgia, but doubted that bamboos could grow that far north. He was proved wrong. The land on which these bamboos were found to be growing was eventually purchased in 1919 and given to the US Department of Agriculture for bamboo research on the numerous new bamboo species that he, Lathrop and Meyer introduced.

The influential scientist Floyd McClure graduated in 1919 and studied extensively in Southern China for the US Bureau of Plant Introductions, sending back many plants to North America. When he left China in 1941 he published over 40 new species, and continued writing, researching and lecturing on bamboos. During the first half of the twentieth century, many new temperate species were introduced into North America at a time when there was no interest elsewhere. These were particularly plants from the genus *Phyllostachys*. Many species now common in North America are only just being cultivated in Europe. During this period many papers were published by Robert A. Young and E. A. McIlhenny; both were almost obsessed with bamboo and their works were very influential.

After the 1940s, interest also died in North America because of the rising importance of plastics, and the abandonment of bamboo for paper-making projects. Politics stifled any interest in oriental gardening during and after the Second World War – Pearl Harbour did nothing to encourage interest in anything Japanese, and the communist takeover in China closed the doors on that country until the Cultural Revolution had run its course (1990s). For over 25 years the progress made over the previous century was forgotten. It was not until the publication of two recent classics in the mid-1960s that some slight interest returned. A.H. Lawson printed *Bamboos – A Guide To Their Cultivation in Temperate Climates*, based on his experiences tending the bamboo collection established by his employer Dr Mutch at Pitt White (p.142), and Floyd McClure issued *Bamboos – A Fresh Perspective* the long awaited results of his lifetime work in North America. Both Lawson's and McClure's books are just as useful today as when they were issued, but in need of updating.

Australia and New Zealand

Climates in the southern hemisphere are very suitable for the commercial exploitation of tropical bamboos, and at least six species are native to northern Australia. In addition Arthur Phillip is known to have brought bamboo from Cape Town in 1788 as one of a number of useful plants introduced to support the new colony. In spite of this early interest very little was heard of bamboos again until Baron Ferdinand von Mueller, as head of Melbourne Botanic Gardens, wrote several papers between 1871 and 1881 extolling their commercial virtues. There was little response to his efforts and by the turn of the century Australia still had probably less than thirty species in cultivation, and the majority of those were grown for horticultural purposes. Traditionally the governments of Australia and New Zealand have been wary of allowing plant importations for fear of exotic species running wild or the introduction of pests and diseases. It was not until very recently that significant numbers of imports were made, and Australia now records nearly 200 species, including many tropical plants with commercial potential.

Although New Zealand has no native bamboos, and no early imports were made, practical research into the usefulness of bamboo species has been undertaken. Species such as *Pleioblastus hindsii*, *Pseudosasa japonica* and some species of *Phyllostachys* have beeen used for controlling erosion and for minor construction purposes. Also largely because of the efforts since 1921 of the late Harald Isaachsen, a large collection of species was established on good land at Oratia, north of Auckland. His son John Isaachsen has continued this work expanding the family firm Isaachsen & Co, and it now supplies huge clumps in a large number of species to landscape projects, gardeners, public works and commercial growers. It also supplies small plants and dried culms for the retail trade.

CONCLUSION

The resurgence of interest in the northern hemisphere in the 1960s sparked the embers that had long gone cold. But they continued only to smoulder for the next 25 years. Then suddenly there was spontaneous combustion worldwide, with a rapid increase in interest in bamboos: societies were formed, good sources of supply were established, and scientific and horticultural establishments revitalized their collections and embarked upon programmes of study.

Even though this is very recent history, it is difficult to understand. Quite what blew on the smouldering embers is unclear. It could be pinned on the rising general interest in the elegance of grasses, but the worldwide nature of the enthusiasm and the all-embracing interest, not just horticultural, suggests that care for the environment, and an interest in permaculture, endangered species and traditional skills were probably equal catalysts. The rehabilitation of Japan into a country with an economy and culture to revere, and the emergence of the sleepy dragon, China, as a dynamic new geo-political player, must also have had some influence.

History will judge today's activities and put everything in perspective more precisely than I am able to, so we will leave others to comment on the amazing progress that is now being made. The questions of yesterday have largely been answered, only to be replaced by yet more questions: such is the way of science. Nomenclature is at long last reaching some sort of stability, but will probably always be subject to flux. Propagation techniques have been improved and as a result only the rare plants are difficult to obtain. A good selection of basic species and forms is readily available to satisfy the average gardener, and thriving societies are to be found in most countries.

Bamboos belong to a world of superlatives and conundrums. They are the fastest-growing terrestrial plant, they are grasses the size of trees, they have the strength of steel and seem to live forever. How do they coordinate simultaneous flowering over large areas of the globe after intervals that we are unable to comprehend? Science requires a continuous study over successive plant generations, so in the age of computers that can solve the most advanced problems in seconds, some of these questions will still remain unresolved long after we are gone.

2
BOTANY

Woody bamboos (as distinct from the tropical herbaceous species that are not covered by this book) have a number of distinctive features compared to other grasses. These features can be summarised as follows: lignified (woody) culms, two phases of shoot growth, rapid shoot elongation, complex culm branching and cyclical gregarious flowering. They also differ considerably from most other garden plants in many interesting ways.

Botanical science divides the angiosperms, or advanced flowering plants, into dicotyledons and monocotyledons, based on whether they produce a pair of seed leaves (dicotyledon) or a single seed leaf (monocotyledon). Broad-leaved trees (as opposed to conifers) and shrubs are dicotyledons and bamboos are monocotyledons. It is essential to understand something of the difference in the growth pattern between these two divisions in order to be able successfully to grow and propagate bamboos.

Most shrubs and trees of equivalent stature to bamboos consist of a deep, branched root system, balancing and anchoring the top growth, which is also usually a randomly branched structure, as well as providing nutrients for it. All growth is produced in the cambium, a single-cell veneer of living material on the outside of the central core of dead wood. The cambium produces the internal plumbing, the xylem, for transferring the sap from the roots to the leaves. When it dies, the xylem contributes to the central core, forming annual rings. On its outer side, the cambium produces the phloem, which is a system similar to the xylem for transferring the nutrients formed in the leaves to the rest of the plant. The phloem later forms the protective bark and is shed. These plants get gradually larger not only through increasing their diameter but also by annual extensions to the growing points on the roots and top growth. Most have a root system that expands to mirror the parts above ground.

Phyllostachys aureosulcata 'Spectabilis' is a useful form that combines hardiness with colourful culms and rapid growth.

Bamboos have a structure that is all living material. They do not possess a cambium layer and do not, therefore, have the ability to increase the diameter of the roots, stems (culms) or branches once they are formed. In order to increase in size as the plant gains strength, bamboos increase the number of branches (some genera) and leaves (all genera). They also have a system of constantly expanding rhizomes supporting an increasing number of culms (canes – the term 'cane' is normally reserved for dried dead culms). The new rhizomes and culms are able to adjust their diameter and length, up or down, to make optimum growth in the changing environmental conditions. Far from being a disadvantage, this system allows the plant to adjust rapidly to improved conditions, such as an opening forming in a woodland canopy. It also enables it to reduce its size without resorting to die-back should the conditions worsen.

INTERNODES

A common structural system of tubular internodes, sealed at each end by a solid diaphragm, or node, is used throughout the rhizomes, culms and branches, and although these vary in proportion and detail, there is little clear botanic difference between zones. The node itself comprises a horizontal or near-horizontal scar resulting from the culm sheath attachment. Above

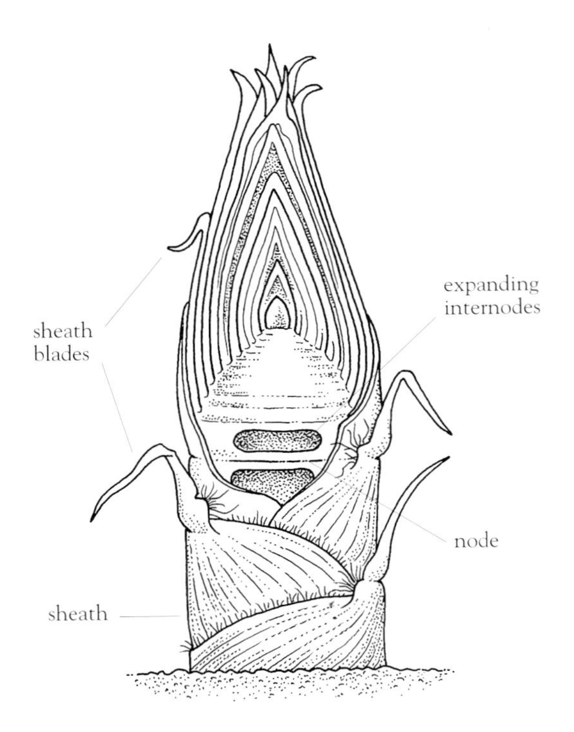

Culm bud in the early stages of extension. The diameter of the culm is already fixed when the internodes expand.

is a supernodal ridge of varying prominence and between these two features is the branch bud, or buds. Internally the diaphragm closes the culm sections at this point. It is usually comparatively thin and may be incomplete with a circular hole near the centre.

The genus *Chusquea* has culms that are solid throughout, and parts of other bamboos are also sometimes solid, such as the base internodes of each section, i.e. rhizome, culm and branch. The ratio of internode length, wall thickness and diameter varies mainly to suit the loads imposed on the above-ground parts, and presumably the botanic requirements on the unstressed parts below ground.

The culms of the large tropical species develop to a height of up to 30m (100ft) or more, with a diameter of 20cm (8in), in about three months, the majority of growth taking place over just a few weeks – an amazing rate of growth, unequalled by any other terrestrial plant. This means that an enormous biomass must be produced by the culms, which at this stage are without leaves and still being conceived. It is clear that the development of the new culms draws heavily upon the food reserves of the established parts of the plant, even if the nearest culm is some distance away and the plant covers a large area. In addition to their work as a rooting system, the interconnecting rhizomes can be considered as pipelines for the transfer of nutrients from one part of a very extended plant to another. Temperate species are considerably smaller, but proportionally very similar, and this does not detract from the wonder of the systems that have evolved to achieve this remarkable feat, for even these culms can grow over 35cm (15in) in a day. When the culm is fully extended no further growth takes place in subsequent years.

NODES

The node areas are the important centres of activity in bamboos. It is at the node area only that buds are found for forming culms, branches, roots and rhizomes. Each of these buds is a complete preformed compressed culm or branch. When it is ready to expand, the bud's potential has already been determined by the food reserves available and the suitability of the growing conditions. Its extension takes place from the node (not at a growing point as in dicotyledons). In the bud stage all the compressed internode within the bud is meristem (the area of active cell division), but as the upper cells of each internode rapidly expand and differentiate into various cell types, meristem activity is progressively restricted to those cells lower down the internode, until the cells throughout cease to divide and expand and the internode is fully extended. This forces each diaphragm apart lengthways, so producing growth rates that are inconceivable in all other plants.

SHEATHS

The sheaths that protect the soft new growth until it has had time to strengthen are fixed at the nodes and produce the distinctive circumferential groove or scar. They are very important to growth. If they are removed prematurely or damaged, the associated internode will be stunted. The reasons for this are unclear; it could be that the sheaths control the growth hormones, but it could also be that the parts of the culm that are accidentally exposed harden prematurely.

All nodes, wherever they are located, have a sheath. This is often jettisoned when its purpose is over, or sometimes it is persistent but dead. Each section, or axis, be it rhizome, culm or branch, starts from a bud on the previous axis. This bud comprises the full complement of nodes tightly compressed with each sheath overlapping like the layers of plywood, but free to slide over one another as the internode lengthens beneath. Growth takes place from the bottom node upwards in sequence. One is reminded of the sections of a car aerial. As the internode expands to its final length, the sheath, which is amazingly strong for such a thin structure, hugs the new culm, rhizome or branch very firmly, imparting temporary support and protection against insects in this vulnerable stage. The diameter of the culm is fixed by the diameter of the bud or shoot, and does not increase at the vertical growth stage or in future years. The sheath has a secondary protective role, for inside its base forms the new bud for the next axis, complete with its own protective sheaths.

Although essentially the same structure, the sheaths are very different, depending on their position on the plant, and have refinements to assist and reflect different roles. The need to force their way through hard earth and to negotiate obstacles is clearly the main function of those on the rhizome, so they are thick and strong with very few appendages. They are also short to reflect the short node spacing. Culm sheaths need to be less strong but have more provision to prevent damage to the culm. They also show a slow progressive change from the bottom to the top of the culm.

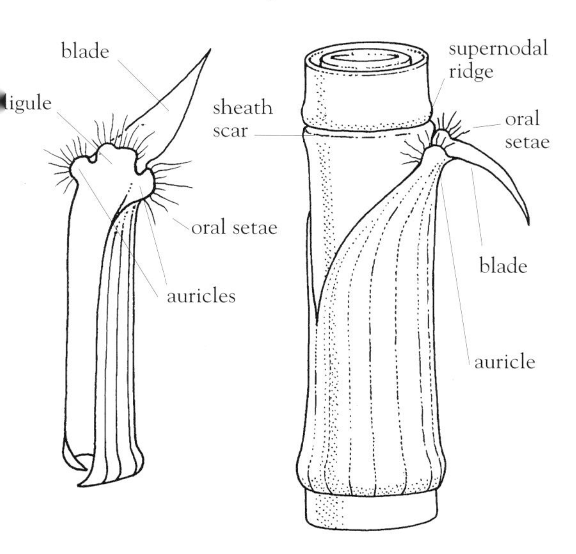

Sheaths: inside (left) and showing the expanding culm protected inside the sheath (right).

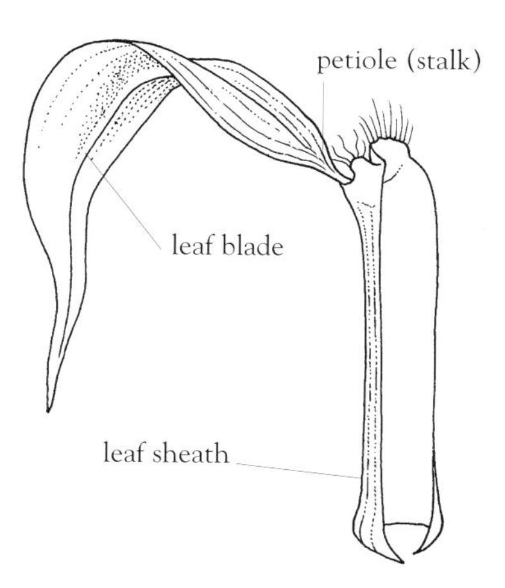

Inside of a leaf sheath showing identical structure to the culm sheath (below) but with the addition of the petiole.

Sheaths appear different from species to species with the additional features varying, too. These differences, plus their texture and colour, are very important in identifying plants, particularly in such genera as *Phyllostachys* where the species are botanically very similar and where other features are unreliable, such as on immature specimens. Sheaths collected at mid-culm position are used for identification purposes but eye level is accurate except for very large or small plants.

SHEATH BLADES AND LEAF BLADES

Culm sheaths have a blade at the top. Those on sheaths growing on the lower part of the culm are only a small triangular extension, but they enlarge with each successive sheath until those at the top resemble a true leaf with chlorophyll. They may even have variegations if the plant is a variegated form. Usually the blade is angled away from the culm or even reflexed downwards on the upper sheaths. At the base of the blade, where it joins the sheath, there is often a short extension to the sheath on the inside of the blade. It projects beyond the base of the blade and hugs the culm profile tightly. This is the ligule and it presumably acts as a seal to prevent the ingress of water, insects and other foreign bodies that are directed to this vulnerable spot by the blade. There may be ear-like projections (auricles), attached either side of the ligule; sometimes there is also a cluster of kinked hairs (oral setae), at the apex of the auricles. When the auricles are absent the oral setae can be attached directly to the ends of the

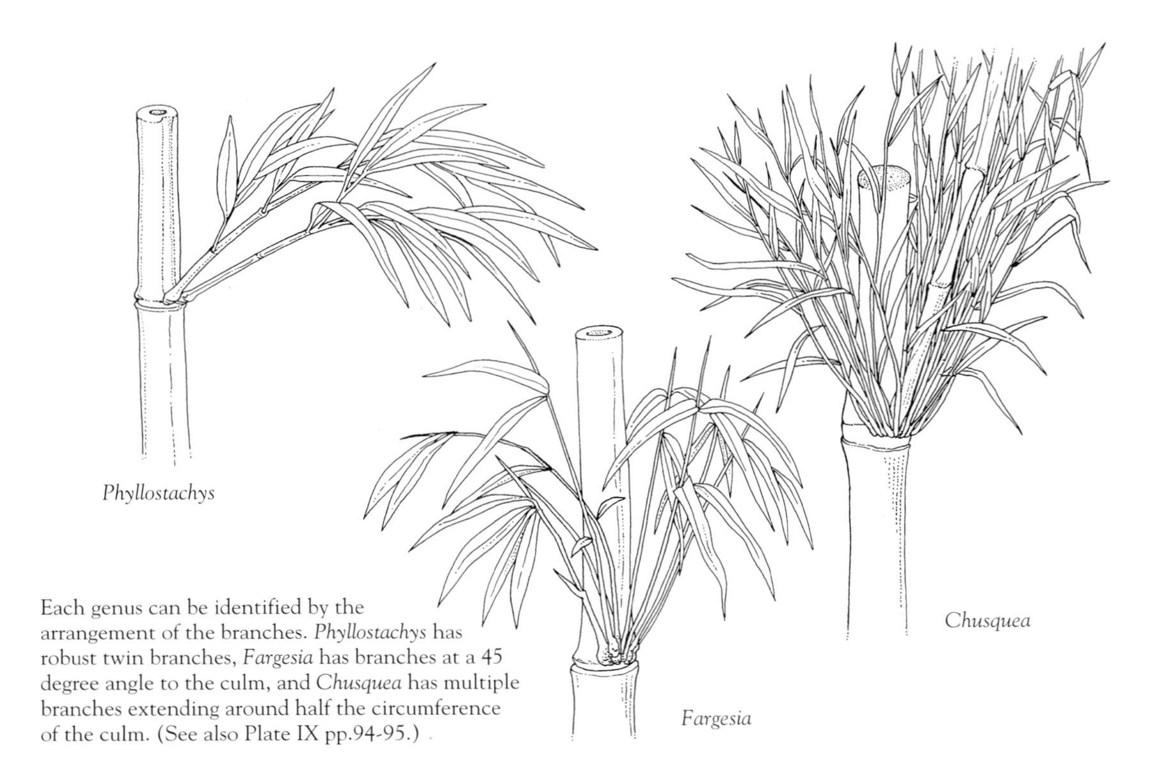

Each genus can be identified by the arrangement of the branches. *Phyllostachys* has robust twin branches, *Fargesia* has branches at a 45 degree angle to the culm, and *Chusquea* has multiple branches extending around half the circumference of the culm. (See also Plate IX pp.94-95.)

ligule. These can be seen holding beads of moisture after rain or a heavy dew, and so they probably assist the function of the ligule.

Culm sheaths protect the branches and there are also branch sheaths that protect any further branching. These are scaled-down and generalized versions of the culm sheaths. Branches are formed like small culms, but the branch formations are a very important feature in the identification of a genus, from the single branches of the *Sasa* group to the multiple ramified branches of the *Himalayacalamus* species.

Although it is not immediately obvious because of their considerably altered proportions, the true leaves or leaf blades of the plant are also based on the same arrangement as the culm sheath blades, as the assistance of a small hand lens will quickly show. The leaf blades are very much larger in proportion to their sheath than even the largest culm blades. They are also connected to the sheath by a stalk (pseudopetiole), which gives them their characteristic obtuse shape at the base. The leaf sheaths and leaf blades are also persistent as their function is ongoing, whereas those on the rhizomes and culms are transient.

The leaf stalks are one of the distinguishing features of bamboos; other grasses generally have wrap-around leaves at the basal connection. Leaf stalks allow bamboos to shed leaves when they are old or under stress. It also allows them to move with the wind and to adjust their orientation to suit the lighting conditions.

Leaves

Bamboo leaves are unlike the leaves of most other large temperate evergreen plants, which have thick leathery leaves with a glossy cuticle surface for winter protection. Having evolved from the more delicate structure of persistent tropical leaves, most hardy bamboos have a system of cross veining, which is thought to give winter protection. Hold a leaf up to the light and you will observe the central vein, or extension of the stalk, running the length of the leaf. Parallel to this is a series of finer veins covering the leaf surface. At right-angles to these are a series of cross veins, dividing the leaf into small rectangles or tessellations. On some semi-hardy bamboos these tessellations are absent although the presence of them does not necessarily indicate that the plant can withstand temperatures below freezing.

Bamboo-like grasses are distinguished from bamboos by their lack of complex branching and leaf stalks. Here, the two larger grasses are *Arundo donax* 'Variegata' (left), and *Miscanthus sinensis*. The small-leaved grass is a species of *Lasciacis*.

ROOT SYSTEMS

Bamboos usually have either a clumping (pachymorph) or running (leptomorph) root system. The characteristics of bamboo seedlings hark back to their tropical ancestry: they are generally less hardy, soft and with a clumping rootstock.

PACHYMORPH ROOT SYSTEMS

A clumping or pachymorph rhizome usually has short segments between the nodes and quickly turns up to form a culm. Buds form on this rhizome and produce only further rhizomes. These, in turn, quickly change to form additional culms. This produces a compact well-defined root system that expands continuously. Sometimes, particularly in heavy shade, the parts below ground can have more distance between culms, producing an open structure, but this can in no way be called a running rootstock as all rhizomes terminate in a culm. Normally this type of rootstock is larger in diameter than the culm it produces. The junction between the old rhizome and the new is called the neck. It is small in diameter, solid, is not able to form buds or roots and is usually short. Among the temperate genera, *Yushania* has a greatly extended neck – I have seen it 2m (6ft) long in *Y. anceps*, where it bridged between pockets of soil on a rocky outcrop. From the above-ground growth, this type of neck could make the species be mistaken for one with a running rootstock.

The extended neck is useless for propagation purposes: an inspection of the rhizome between culms will show a clean series of small-diameter internodes devoid of buds or roots, as distinct from a running rhizome, which has roots, buds and often redundant sheaths at every node. The neck's terminal growth is also distinctive. Close-packed culms form around these well-spaced culms by normal tillering (production of shoots from the base of a culm). All pachymorph species are safe, in even a very small garden, but *Yushania* species should be introduced with caution or restricted to medium-sized or large gardens.

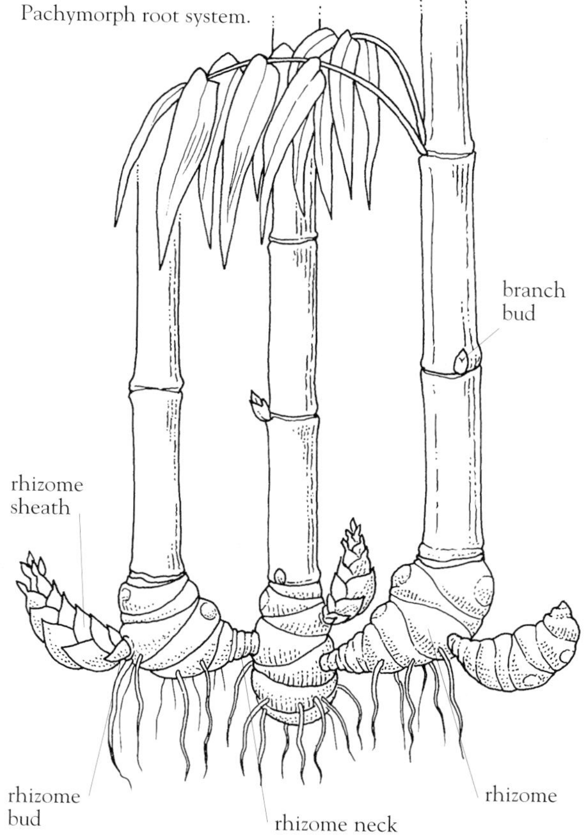

Pachymorph root system.

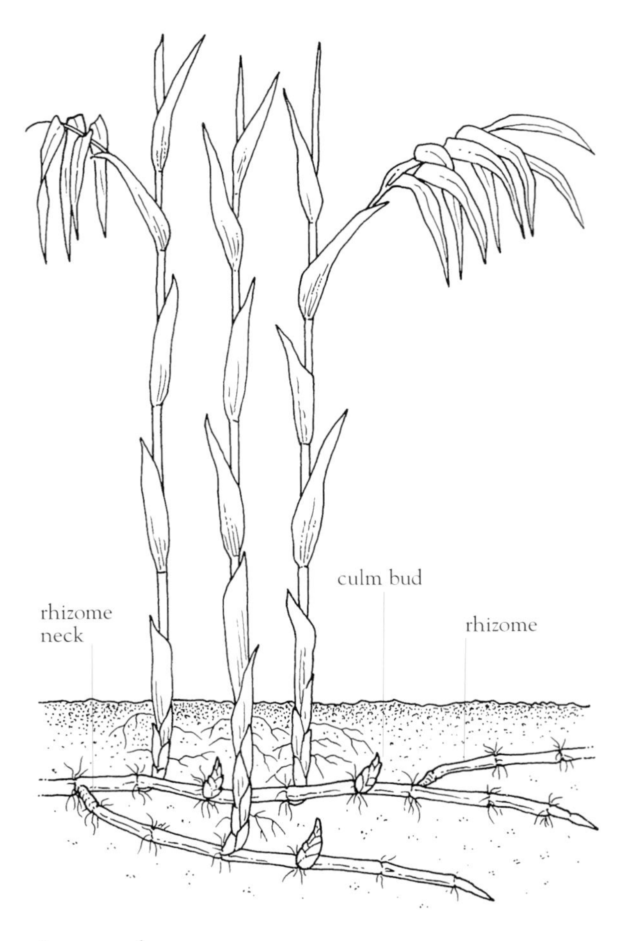

Leptomorph root system.

LEPTOMORPH ROOT SYSTEMS

Many temperate bamboos have evolved a running, or leptomorph, root system where the rhizome is more or less like a horizontal culm, having roots and side buds at each node. It is also similar to a culm in that, after a period of extension, it usually reaches a point where further growth ceases, normally coinciding with winter or a period of drought. Alternatively, the rhizomes may turn gradually upwards and eventually evolve into shallow-angled culms. Normally culms are produced from the rhizome buds, which turn quickly upwards via a narrow culm neck.

Although there are roots on the bottom of the culm there are no rhizome buds, so when propagating a leptomorph species it is essential to include a section of rhizome that has active buds. If not, the section will root and be healthy, but will age and eventually die without producing any new culms. On all but the small species, this type of rhizome is smaller than the culms it produces.

The rooted culm base can change either distinctly from a rooted section to the smooth above-ground culm, or gradually between the two. This varies between species and also between different plants of the same species. It is essentially the same segmented structure, however, and the plant will not suffer if the soil level varies from that at which it was formed. Sometimes the culm base expands considerably in diameter as it enters the soil, probably to enable it to produce a greater number of roots.

After the rhizome has run out of vigour and ceased to grow forward, future rhizome activity is taken over by new rhizomes. These are formed from the rhizome side buds, which have the usual neck feature. In this way the bamboo can colonize large areas of fresh soil and has botanically achieved a distinct advantage over its pachymorph forebears. Often many more buds are present than are needed. Many stay permanently dormant, although division of the root or damage to the new growth will usually activate some of these into either new rhizomes or new culms. Remember that young leptomorph plants, or plants with insufficient vigour, start life with a clumping rootstock. If they are given cool or poor growing conditions they may remain in this juvenile state. Often the transition between the juvenile state and the true leptomorph rhizome follows a period when each culm emerges in a slow curve from the end of extending rhizomes.

A few leptomorph genera, for example *Chimonobambusa* and *Pleioblastus*, have the ability to combine the advantages of both rhizome systems and are able to produce new culms from the base of other culms. Mature plants of this type appear as bunches of culms with wide spaces between.

INTERNAL STRUCTURE

A study of the cross-section of wood-bearing trees of a reasonable age shows the annular rings forming the majority of the central area, which now forms no functional part of the living plant, but merely acts as a support system. A similar study of a bamboo reveals that it is all living material. A section across the culm wall has a decorative pattern of dark spots on a light background. These spots become smaller but more concentrated towards the outer edge. An enlargement of one of the spots shows that it comprises a number of tubes, both xylem and phloem, in vascular bundles incorporating the strengthening fibres in characteristic patterns. These patterns vary between species, and gradually change according to their position within the culm wall and with culm height, or their positions in the rhizomes and branches. Most temperate species have one central vascular system per bundle with the fibres arranged in surrounding sheaths.

The fibres, so essential for the culm strength, represent about 60–70 percent of the weight of the culm or 40 percent of its volume, and are generally longer than those found in hardwoods. The shortest fibres are always towards the node. They spread from their parallel arrangement in the internode wall to incorporate the node diaphragm and branches. This arrangement makes the node the weakest point and often a point of fracture on immature culms, but it also allows a dead segment to break cleanly at the diaphragm. The proportion of fibres increases noticeably towards the outer walls, where the bending stresses are greatest. In addition these outer fibres are of laminated construction with improved structural properties.

The vascular and structural fibres are spaced in a matrix of cells, or parenchyma, in a very similar way to carbon fibre. The parenchyma quickly becomes lignified as the culm expands, so that, soon after the culm has reached full height, the sheaths that have provided temporary support are no longer required. They die and are often shed. On the inner culm skin there are numerous layers of greatly lignified cells that provide an impervious layer. The outer culm skin is very hard, waterproof and glossy in most species. It consists of two epidermal layers of cells that have a very high silicon content and cutinized walls. The various culm cells have to remain alive for at least 10 years, or the life span of one culm, without the formation of any new tissues. For this reason it is best to use young material for propagation purposes as, often, older culms have lost the ability to produce new cells.

Bamboos do not find a need to produce toxins in the culm material to resist insect attack as some trees do. This is presumably because of the comparatively short life of an individual culm and its very strong surface. The waterproof surfaces of bamboo do not easily absorb

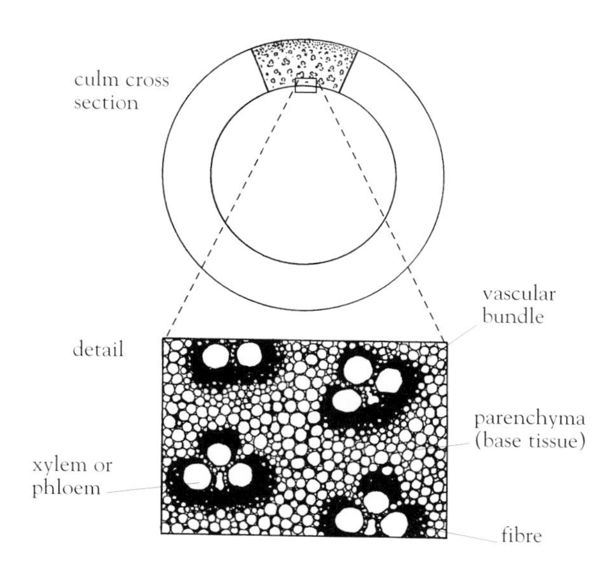

Vascular structure of a typical bamboo culm. The bundles are smaller and more compact towards the outside of the culm.

the preserving compounds that are used on normal wood, so its life as a timber product is often comparatively short.

FLOWERS

Once in a while a bamboo flowers and reminds us that it is not immortal. A whole chapter could be written on what is not understood about the flowering of bamboos: only a few sentences are needed to record what can be explained. Flowering can be partial or complete. It can be more or less in unison, including the majority of plants worldwide whether they are wild, cultivated, far from home, growing in a greenhouse or in complete isolation. Sometimes so much seed is set that flour is made from it in some parts of the world, but at other times little or even no seed is produced. Some species have never been known to flower, some never to set seed. In the Far East, over many generations, clones have been selected of useful species that never seem to flower. The implications of this are that the plants do not age, otherwise they would have reached senility and died without ever reproducing naturally. Obviously all living things age: it is just that bamboos live in a much longer timescale than other more familiar living things; or perhaps it is that part of the propagating technique resets the time clock.

PLATE II

Bamboo rhizomes

Scale approximately half lifesize

Pachymorph rhizome (*Fargesia dracocephela*)

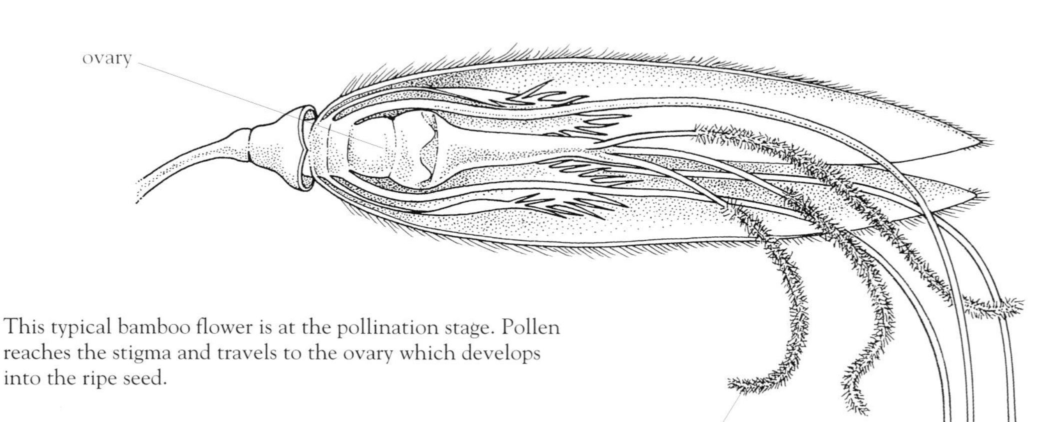

This typical bamboo flower is at the pollination stage. Pollen reaches the stigma and travels to the ovary which develops into the ripe seed.

To explain the unexplainable it is first necessary to forget all the generally held beliefs that have evolved. To understand the behaviour of these plants for certain we need to be able to sow seeds and raise a generation under observation. We need then to watch them flower, raise a new generation, which then flowers, and so on until a picture emerges. As bamboos, in most instances, have lifespans that are many years more than ours, this is plainly impossible. Lifespan figures published in books such as A.H. Lawson's '*Bamboos*' and C.S. Chao's *A Guide to Bamboos Grown in Britain*, are nearly always from random records based on a multitude of clones; the true figures could be much longer.

The phenomenon of synchronized flowering is hard to explain. Some experts hypothesize that all cultivated plants come from one clone, but this is over-simplistic and such arguments do not take into consideration the huge quantity of plants that were imported during the last century. Plants came from different countries, under different names, some ethnic and some incorrect, and some came from seed. It would be even more incredible to think that every specimen of every species just happened to be a clone than that bamboos have evolved to find their own way of synchronizing flowering. Also plants that have been grown from an imported batch of seed often flower in unison, such as *Chusquea culeou* at the time of writing, and these are definitely not a single clone.

What is known is that it is essential for a wind-pollinated plant to flower in unison with a neighbour of a different clone if it is to avoid self-pollination. We also know that when a plant is relying on random pollination, complete flowering is bound to be more successful than just producing a few flowers.

Complete flowering

When a bamboo flowers completely, most of the leaves are usually replaced by flowers, transpiration is largely interrupted, and this triggers natural responses that hasten the aging of the culm. The plant lives off its food reserves, which would otherwise be used for the next generation of culms, and the culms eventually die. If there are no food reserves left, they are not replaced, or they are replaced temporarily by small culms, which in turn flower and die. Death seems to be the normal end result when a pachymorph species flowers completely probably because its short rhizomes are integral with its culm. Leptomorph plants, if mature or growing strongly, usually have sufficient food reserves left, presumably in the rhizome system, to start a new generation. New shoots arise from the base. They are juvenile in form, with small soft culms and large leaves, and over a few years they again become mature plants. This is obviously a time of great cell activity, for frequently selected forms revert to the type plant, or sometimes new different forms arise. It can be a time of anxiety tinged with hope for the horticulturist. Some of our best garden forms of *Phyllostachys* have developed after flowering, often from other forms, and often the same form appears independently in different parts of the world.

Yushania, with their extended pachymorph root system, have also been known to arise again from the dead, often after an interval of a few years. The famous clone of *Yushania anceps* from Pitt White (p.137) flowered and died in unison with most other plants of this species a few years ago. Many seeds were collected, raised and sold under the name of the parent. Then a few years later, the original plant, which fortunately had not been dug up, started to grow again. The new growth was unlike seedling growth in that it was immediately large and spaced out, rather than being many compact tiny culms getting larger as the plant matured, so there was no doubt that this was regeneration after several years' dormancy. Other similar reports are known where the apparently dead roots of this species were not destroyed by the gardener. Perhaps, therefore, we should also preserve the root system of all of our flowering pachymorph plants in the slim hope that some others will regrow.

Partial flowering

There are many instances of partial flowering over several years. Here some culms remain unaffected and live on after the phase is over. More frequent is the isolated flowers occasionally seen on most plants. Some species have been known to flower only in this way. Both could be the precursor to complete flowering. When flowers are seen on a plant, unless it is complete flowering in a pachymorph, it is not always the disaster that we have come to believe. It is usually an opportunity to collect seed, to raise a new generation that will not normally flower again in our lifetime, and to look for new forms or seedlings better adapted to the location.

SEEDS

Many people record that bamboos rarely set seeds, or do not set seed at all. This is sometimes true for a few species, during partial flowering and for species growing in conditions far removed from those that they enjoy naturally. However, in my experience, it is very rare that bamboos flowering in earnest do not set some seed during one of the years of their flowering cycle. Usually complete flowering is preceded by partial flowering and often this does not set seed. Similarly the reduced flowering after the peak years rarely produces seed. But at the peak of flowering, reports of this type are usually based on examination at the wrong time of the year or failure to find the seeds that are there.

Some seeds, such as those of *Pleioblastus* or *Pseudosasa*, are easily seen as they are large and prominent, but the majority need to be searched out and even then some are difficult to distinguish from the dead flower parts. As the bamboos have evolved from tropical plants, they drop their seed as soon as it is ripe, unlike many temperate plants, so, to find seed that is in the best condition for germination, it is essential to examine the plant at the correct time of the year, and to do this a basic knowledge of the flower parts and some understanding of the flower cycle is necessary.

Fresh flowers are relatively colourful and may show thread-like yellow extensions, which are the anthers or pollen sacks. At this stage, seeds have not yet begun to develop. Seeds collected prematurely will not be fully developed and will be difficult to germinate and grow on, but if the flowers are brown and dried they have probably already dropped their seeds. There is no easy solution to this other than by observation, but seed will probably be ripe 4–5 months after the anthers are seen. This will vary with climate, but generally the mountain species will start to flower in late winter or spring and ripen by midsummer, while the *Phyllostachys* flower when the weather gets warmer and the seed ripens as autumn approaches.

If the plant is your own, observe it frequently. When the flowers are turning brown, feel the base of each floret to see if there is a hard object (the seed) within. Do not be too enthusiastic over this. If you feel nothing, you may be destroying developing seeds. The hard object may sometimes be a collection of dead material, but a detailed examination will distinguish between this and seed. If it is a seed, and it is released without

PLATE III

Bamboo flowers

Scale approximately half lifesize

Pleioblastus simonii
'Variegata' (simple)

Pseudosasa japonica
(simple)

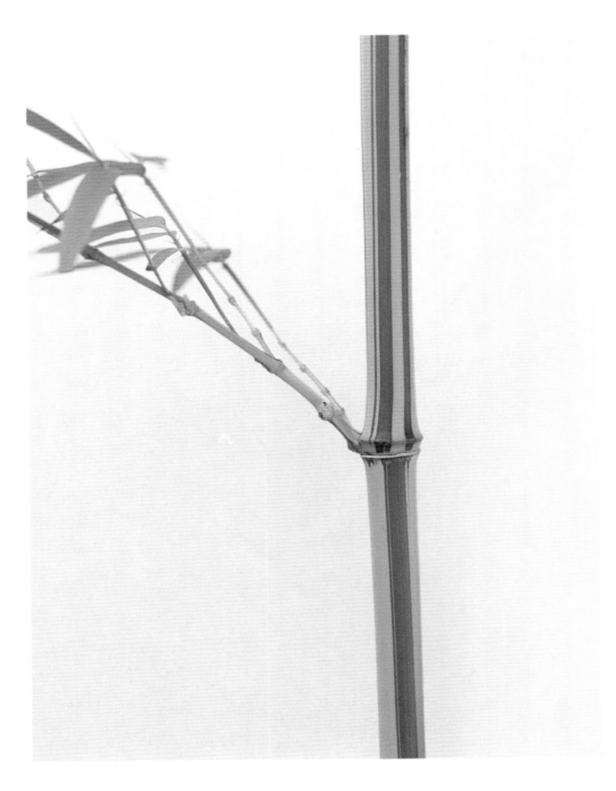

Phyllostachys vivax 'Aureocaulis' is one of the finest bamboos with coloured culms. It can grow to a very large size and is one of the more hardy species.

any persuasion, then you need to collect what you can from all the flowers as soon as possible. If you have to use a little force to release it from its stalk, then it is unripe and other flowers should be left a while. An alternative is to spread an old sheet on the ground to catch the seeds as they fall naturally, but this can only be done if there are no plants underneath, and, although you will collect perfectly ripe seeds, it is likely to become a picnic table for all the local wildlife. Another method of ensuring that you obtain some seed is to wait until the flowers start to turn brown and then to cut the complete culms into suitable sections, with all the branches in place. Store them loosely in an open-topped plastic bag until the seeds are ready to fall naturally. Some will drop into the bag but most are held in the dead flowers and need searching out. The sustenance within the dying culm should be sufficient to ripen the seeds, provided it is not cut too early. This is a useful method if the plant is not yours and you are not able to visit frequently. It is far better than picking immature seeds, however tempting.

Many of the flowers in a spikelet are rudimentary and sterile, so this reduces the number of seed that are set. On one occasion I collected over 2kg (4½lb) of seed from a plant, but often it is more like two dozen seeds. If you have decided to grow plants this way, you will need plenty of patience or you will be joining those who maintain that bamboos rarely set seed.

The seed of bamboo normally loses its vitality very quickly, and any seed collected should be distributed or sown without any delay. If the parent is growing well and the seed is collected when ripe and sown quickly, viability is high and there are normally few problems with germination. Germination should take place between a few days and a few weeks after sowing, but a few genera, such as *Indocalamus* have a built-in delay of at least a year. Some Himalayan or other mountain species, if allowed to enter the dormant stage by any undue delay, seem to require a few weeks of cold to break this dormancy. This is probably their natural life-cycle so it is often best to keep these species in a domestic refrigerator for a few weeks before sowing (p.66).

Seedlings

Seedlings first send up a single shoot. It can hardly be called a culm. After a while, the exact timing depending on the vigour of the seedling and the species, other larger culms develop from the base in pachymorph style. After a year the young plant should be 5–15cm (2–6in) high and comprise several small, soft culms gradually increasing in size and strength. Species that are semi-hardy will probably be unable to withstand frost until the culms are hard and woody, and should be overwintered in a greenhouse until they are strong, mature plants. Depending on growing conditions, it can be several years before leptomorph species start to develop their adult rhizome system and lose their large juvenile foliage and vulnerability.

VARIEGATION

The occurrence of variegated forms is fascinating, and involves characteristics not often seen in other plants because of the bamboo's unusual form. The first major difference is the formation of chlorophyll in the culm, which, associated with a clean, often glossy surface, is at least as colourful as the leaves. In addition, the culms have various boundaries, such as at a node or a sulcus (the groove formed on some species above the branch bud), where changes to the pattern of variegation can take place.

Variegations in bamboos are the result of a chimera, which is a type not reliably transmitted via seeds. All variegated bamboo plants comprise a combination of two types of tissue, the mutant tissue that has lost its chlorophyll and appears white or yellow, and the normal green part that provides the nutrients. To be stable and to be consistently reproduced, both types of tissue need to be present in the growing point. In a chimera this is achieved by one layer overlaying another.

There are usually up to three external ring-shaped cylinders of tissue that are liable to variegation effects. Where the outer layer is mutant and the second layer is green, this shows up as a yellow culm only, but where the lower layer is exposed, in a species with a sulcus, for instance, this groove is green. *Phyllostachys bambusiodes* 'Castillonis' is an example of a yellow-culmed bamboo with green grooves. Sometimes this mutant layer is unstable and is disturbed by the complexities of the cells at the nodes, and then random stripes can occur such as are seen in *Phyllostachys vivax* 'Aureocaulis'. Where the top two layers are mutant, all-yellow culms are produced, such as those of *Phyllostachys bambusiodes* 'Holochrysa'. In these the colour is usually richer because of the double layer. A top green layer over a yellow layer gives green plants with a yellow sulcus. Where the top two layers are green, all-green culms are produced, but sometimes these have a third variegated layer hidden, waiting to be exposed at the leaves.

As the system of cylindrical layers transposes into the leaf structure, the neat arrangement breaks down and glimpses of the various layers are revealed, so we often see variegated leaves in combination with coloured culms, or in isolation. Variegated leaves can also sometimes be found on plants that have been stressed some way. This is usually because the winter temperatures have been close to the minimum that the plant can tolerate, but the effect is temporary and should not be confused with a mutant form. Sometimes this stress-related effect produces pure white leaves with no variegations.

In addition to the basic culm colour, some bamboos have an overlying pigmentation, such as the dark shading that can be seen on many green bamboos when they are exposed to the sun. When this pigmentation occurs over the yellow of a variegated culm, it can produce the bright red effect that is seen on *Himalayacalamus falconeri* 'Damarapa', or *Phyllostachys aureosulcata* 'Spectabilis'. The colour varies with the degree of exposure to light. In addition there are brown marks or even the solid black of *Phyllostachys nigra*. The pigmentation can be natural or caused by a virus. Plants can also have culms of a startling grey-blue, as seen in some forms of *Thamnocalamus crassinodus* or in *Himalayacalamus hookerianus*. This is produced by a glaucous thick wax coating.

Sasaella masamuneana 'Albostriata' has striking leaves in the spring, but, unfortunately, it is a rampant species that needs some control and pruning.

3

CULTIVATION

Of all the plants I have grown, none have responded so well to good cultivation as bamboos. Provided that the growing conditions are suitable for a particular species, it will prosper with almost no preparation or aftercare. All that may be needed is to ensure that specimens do not dry out for the first year or two. With some extra attention to planting and good aftercare, your newly acquired bamboo can be up to full height within a few years, compared to ten or more years for those unlucky plants that have to make their own way in the garden. However, people with small gardens may choose to cultivate plants without special preparation or aftercare so that they grow slowly and remain within the proportions of the garden, and so that excessive spread is not a problem. Even here, though, it is desirable to encourage the plants to full size as quickly as possible so that the new planting soon looks mature.

Bamboos are very tolerant of different soil types and growing conditions but it is always advisable to prepare the planting hole well. Loosen the soil, add as much humus as possible to all soil types to assist with moisture retention and drainage as this is very important for the first few years. Animal manure can be added if available, or alternatively, a slow-release fertilizer, such as bonemeal, as bamboos are greedy feeders. Bamboos will not tolerate poor drainage, so very poorly drained soil should be dug deep and drainage material incorporated. However, as they are shallow-rooted this extra work is not necessary in most soils, and the fertilizer should only be incorporated into the top spade depth.

Although effective here, *Pleioblastus chino* is a coarse species and other bamboos are more deserving of this protected site.

PLANTING

Plant your new bamboo as soon as the weather is suitable after winter and as soon as possible, for they always seem to prosper better when in the ground. For those areas with regular weather patterns, plant during a cool rainy season or just before if this can be predicted. For pot-grown specimens, I normally soak the bamboo and then plant it a little deeper than it was in the pot. There is no clear distinction between top growth and roots as there is with other plants and slightly deeper planting can stimulate dormant roots on the basal nodes. It can also protect the root from drought or cold. In soils with good drainage, plant in a shallow depression to assist with water retention and to make sure that any subsequent water drains to the place where it is required. Conversely, on wet sites the bamboo should be planted on a small mound to prevent the build-up of standing water around the roots for the first year or two. Soak the plant in well and, ideally, cover the surrounding area with a mulch of rotted animal manure, leaf mould, bark or any other suitable material, to conserve water and to keep the soil warm.

Plants from pots do not normally need staking, but tall transplanted divisions should be given some support if the culms are kept at their full height. Rather than use a single support, as you would with a tree, a more secure system is to use two or three posts and thread the interconnecting rope between the culms. One or two species, such as *Phyllostachys sulphurea* 'Viridis', have a tendency quickly to put out large culms from a compact base. If this happens, it is wise to stake before the winter winds start, making the newly established plant unstable. Plants sensitive to the wind should be given a screen of windbreaking material if the surrounding plant protection is also immature. They will not prosper, or even survive, if wind blasted. On an exposed site all bamboos, even the indestructible ones, should be given this protection until they have sent out new culms that are adapted to the local conditions.

AFTERCARE

If you fertilize well with a long-lasting fertilizer at the planting stage, little aftercare will be needed, but you must ensure that the plant does not go short of water for the first year at least. This is essential. Consider installing a weeping hose if watering during this period will be a problem.

For maximum growth, watering should be continued until the plant is as tall as you want it to be, and a good mulch of rotted manure added every other year. You should be amazed at the results. If the bamboos are in a situation where they will be required to resist wind, drought or other stresses, then encourage harder growth by reducing or eliminating fertilizer.

FEEDING AND MULCHING

The sections regarding the fertilizing of bamboos recorded in the Japanese work *Nikon Chiku-Fu* (p.21) are most interesting. Under the section relating to the culture of the giant timber bamboo (*Phyllostachys bambusoides*) the author records, 'The dead bodies of dogs, sheep, cats, rats and other animals, the skins, bones and hooves of cattle and horses, are the best for this purpose. Decaying rice and wheat plants, rice and barley bran, and other vegetable matter, ashes, the contents of the dustbin, rotten compost, stable litter, the dung and urine of men and horses, and lime where the soil is not sandy, may all be used.' In another part of the book relating to the culture of that temperamental species *Phyllostachys edulis* he records that fertilization is not necessary in the warmer areas of Japan, but gives a list of fertilizers very similar to those for *Phyllostachys bambusoides* that are essential if large culms are to be produced in the colder regions.

In my small garden, where giant culms would not be welcome, I fertilize once a year as soon as I see the very first signs of growth in the spring. I use a commercial lawn fertilizer that is high in nitrogen. This gets the plants off to a strong start without stimulating large and soft growth. Applications of a high phosphate fertilizer from late summer onwards can be beneficial but take care not to stimulate too many culm buds for next year (p.57) by being over-generous.

Probably more important than fertilizing for an established plant is a good mulch and a build-up of natural leaf litter around the roots. Bamboos seem to love decaying wood, leaf mould and old tree stumps, and there is no danger that they will contract honey fungus or other tree diseases. With their upright form they are, therefore, ideal for disguising a dead tree or a decaying stump. Mulches protect the root system from extremes of temperature and drought, and can be coarse composted bark, garden compost, seasoned manure or natural leaf litter.

Most bamboos detest seaweed as a manure, probably because of their sensitivity to salt, and it is best to avoid compounds derived from seaweed as well. Some are so sensitive to fresh or even composted seaweed that it has even been recommended that it is used for restricting the spread of rampant species.

Normally, if the plants have a good root system and if there is a good natural or applied mulch, little further maintenance is required. Once well established, bamboos are remarkably resilient and will survive short droughts and other occasional natural disasters, although if this happens in the growing season the growth for that year, and to some extent the growth for next year, will be affected.

SILICON

There has been much discussion about the role of silicon in the growth rates of bamboos. Silicon is not a fertilizer, however, but only a hardening material, and it is doubtful if it influences the speed of growth. Silicon is present in every soil, which is presumably the source for bamboos, but it is almost insoluble in water (which is the working fluid that transfers all chemicals within the vascular system). It must be readily available to most plants, even if we do not know how it is transported. If a good leaf litter is allowed to build up, there can be no need for the gardener to be concerned about lack of silicon.

PRUNING

The only essential aftercare required is that young plants have any unsightly growth removed and older clumps have any old or damaged culms removed annu-

× *Hibanobambusa tranquillans* 'Shiroshima' (right) forms a tunnel over a path with *Sasa palmata* 'Nebulosa'.

ally to maintain vigour. Old culms are easily identified by their colour, which is usually brown or at least lacks the bright green, or yellow in the case of variegated culms. Also they will have fewer leaves and probably many dead branches as well. Old culms contain little nutrients and, although you are removing a very small amount of food reserves, the opening up of the centre of the plant to air and light more than compensates. A constantly thinned plant is always healthier and better-looking than one that is left to nature, provided the pruning is undertaken wisely.

The effect of pruning on the rhizome system needs to be understood if the plants are to remain in good condition. With leptomorph species, it is very tempting to remove all the stray growths to preserve appearances or for propagation, but it should always be remembered that older root systems lose the ability to form new rhizomes. If we are not to end up with a geriatric plant, that, although it may be tall, sends out no new culms or rhizomes, then we should allow even the most rampant of species to wander a little at some point around its circumference. The older part of the plant can be gradually removed as it ages. Ideally, this wandering should be planned into the garden design and a new division replanted at the original spot when the plant has wandered too far.

Pruning should take into consideration the growing cycle of the bamboo if it is not to weaken the plant or give disappointing results. Remember that the mature

This distichus form of *Pleioblastus pygmaeus* needs a light location to retain its distinctive leaf formation.

culms are the main reservoir for the starches that primarily support the tremendous surge of spring growth. Clear cutting an old stand (clump), to take the extreme case, will not result in strong vigorous new growth, as is normal with a rose, for instance. The resultant new growth will be short and small, as described under the taking of rhizome cuttings (p.64), because the only reserves that it can call upon is what is left in the rhizomes – most of the food will have been removed along with the mature culms. The best time to prune away old or unsightly culms is when the new culms have reached their full height and have begun to put out branches and leaves of their own, and so not need the support of the rest of the plant, as this is the time when the food reserves are at their lowest. Do not cut out more than a third at any one time, as the older culms also physically support the new growth until they are stronger.

All culms that are removed can be used as garden canes, except for those that are two years old or less, or those that have died on the plant before cutting. Culms become harder as they age, and those three or more years old are reaching their strongest. Cutting the culms when the starch content is at its lowest is not only best for the plant but also for the usefulnesss of the dried culm, particularly for those warmer regions with destructive insects, such as termites, that eat the wood mainly for its starch content. Once cut, clean the culms of all branches and allow them to dry naturally on a flat surface in a cool shed. They will then give several years' service, at least as long as bought canes. Undried culms are easier to bend and work for handicraft purposes but have a short garden life.

Cosmetic pruning

When plants are mature, pruning can be fairly adventurous. Many species look better with their lower branches removed, particularly those with ornamental or coloured culms. The best time to remove these is when they are just forming. At this stage they are easily broken off by hand, which is much quicker and neater than cutting with secateurs.

Mature plants that have formed a grove should have all the small-sized culms removed. Although it is not advisable to walk within the plant at culm-shooting stage, for fear of damaging the new buds, any small-diameter shoots that can be seen to be forming should be removed before they have grown very tall.

Low-growing bamboos nearly always look better for pruning. Most 'dwarf' bamboos can reach 1.5–2m (5–6ft) high in good conditions if left to their own devices and even the very low-growing *Sasa veitchii* and *Shibataea kumasaca* look much better if the new culms are trimmed to make a neat cushion of foliage. Plants designed to make a formal or informal hedge should be pruned at the new culm stage (p.69).

Sometimes culms are produced that hang inconveniently over paths or that flop on the ground and are out of character with the plant. These are sometimes formed from the terminal growth of leptomorph rhizomes, or more usually from culms that have been over-stressed during their tender first few weeks. Permanently damaged culms are best removed if they are causing problems. Those that are just a bit weak can usually be brought upright by some judicious, inconspicuous pruning. It is the weight on the end of the culm that exerts the most leverage on the base. Removing a few internodes from the top is usually all that is needed; as these are small this can be done without destroying the appearance of the culm. If this is still insufficient, remove a proportion of the branches, again starting from the top. *Phyllostachys* species usually have two branches per internode, one large and one small. The removal of one of these, all down the culm, will allow it to move upright and will not be noticed. Which branch is removed will depend upon the severity of the surgery thought necessary.

After many years even well-kept plants can become very wide at the base and lose their elegant proportions. Dividing in half (essentially the same as taking large divisions, p.62), will restore their good looks.

Reviving an old stand

If you inherit, or are asked to revive, a very old stand of bamboo, it can be a very daunting prospect, but it can also be rewarding, so long as you can do the job gradually over a couple of years or so. As mentioned before, the extreme solution of cutting off all the culms is not a good idea. Start by assessing the plant and considering the form that you wish to achieve. Is it suitable to make a grove, hedge or a specimen? It is a fair assumption that, if it is very old, it will have grown very broad at the base and inelegant as a consequence. Identify any essentially lifeless areas that will be open when all dead culms are removed. Then decide if you are going to remove parts of the plant to make its proportions more pleasing. Often two or even three elegant plants can be cut out of one rambling specimen. Groups formed like this can look very impressive.

Once you have decided on what parts of the bamboo to retain, the first, and hardest, stage is to perform crude surgery to the base of the plant, cutting out any sections of root that are not required. To avoid killing your enthusiasm for bamboos for life, invite some strong friends to stay with you for the weekend 'to do a bit of gardening'. The next stage is to remove every dead culm on the remaining pieces. This will open up the clump, enabling you to see what you have and also allowing you to get inside the stand. Follow on from this by removing every old culm. This is best undertaken by pulling to one side all the culms that only have a few leaves and then cutting them out at the base. Heavy-duty loppers are ideal for medium-sized culms (up to about 5cm/2in diameter) but a metal-working saw or very fine disposable woodworking saw is needed for bigger ones. Once you have completed this give the patient a good dose of a general fertilizer to enable it to respond to the better condition in which it now finds itself.

All this work can be undertaken at any time of the year, although if the specimen is fit enough to send up new shoots the growing season should be avoided. Further pruning will be of good culms and should be integrated into the growth cycle. Perhaps the year after the basic surgery you should remove all distorted or truncated culms that remain. There will probably be many of these, a legacy from when the plant was struggling. The final stage is to remove all the lower branches that remain if this is needed for appearance's sake. During the work you will have uncovered a multitude of dead branches. These are victims of the poor light and stagnant air within an overcrowded stand. I always remove these as I go, even if the culm is to be removed later, for it makes the work that much easier. Heavy leather gloves and eye protection are essential during all this work, as it is so easy not to see a foreshortened culm or branch as you stretch into the inner reaches. You will also dislodge years of debris, which will take its effect on your eyes the following day.

After years of starvation of both light and nutrients it will take a while before you see a response to your efforts. Assuming that the plant has been adequately fed and watered – the more the better – it will be a very different specimen in a couple of years, and an imposing specimen in the years after.

TRANSPLANTING

Most bamboos, with the exception of the chusqueas and those without a root ball, can be easily transplanted within the garden. Provided that the ground and air are moist, this can be done at most times of the year, except winter, and in warm regions the height of summer is best avoided. Ensure that a good ball of soil is kept around the roots while the plant is being moved, water it well in its new location for the rest of the year and stake it if necessary. Delicate or small plants can be grown on in a protected site and transferred to their final location when they have made good-sized specimens.

Do not be daunted by the thought of moving large specimens. You will need a reasonable amount of time and an assistant if a mature bamboo is to be moved in one piece. Tie the culms tightly together with a rope. Clear the area around the specimen and dig a vertical trench down one side, making it about 75cm (30in) deep and 30cm (12in) away from the outer culms. On

PLATE IV

Bamboo culm shapes

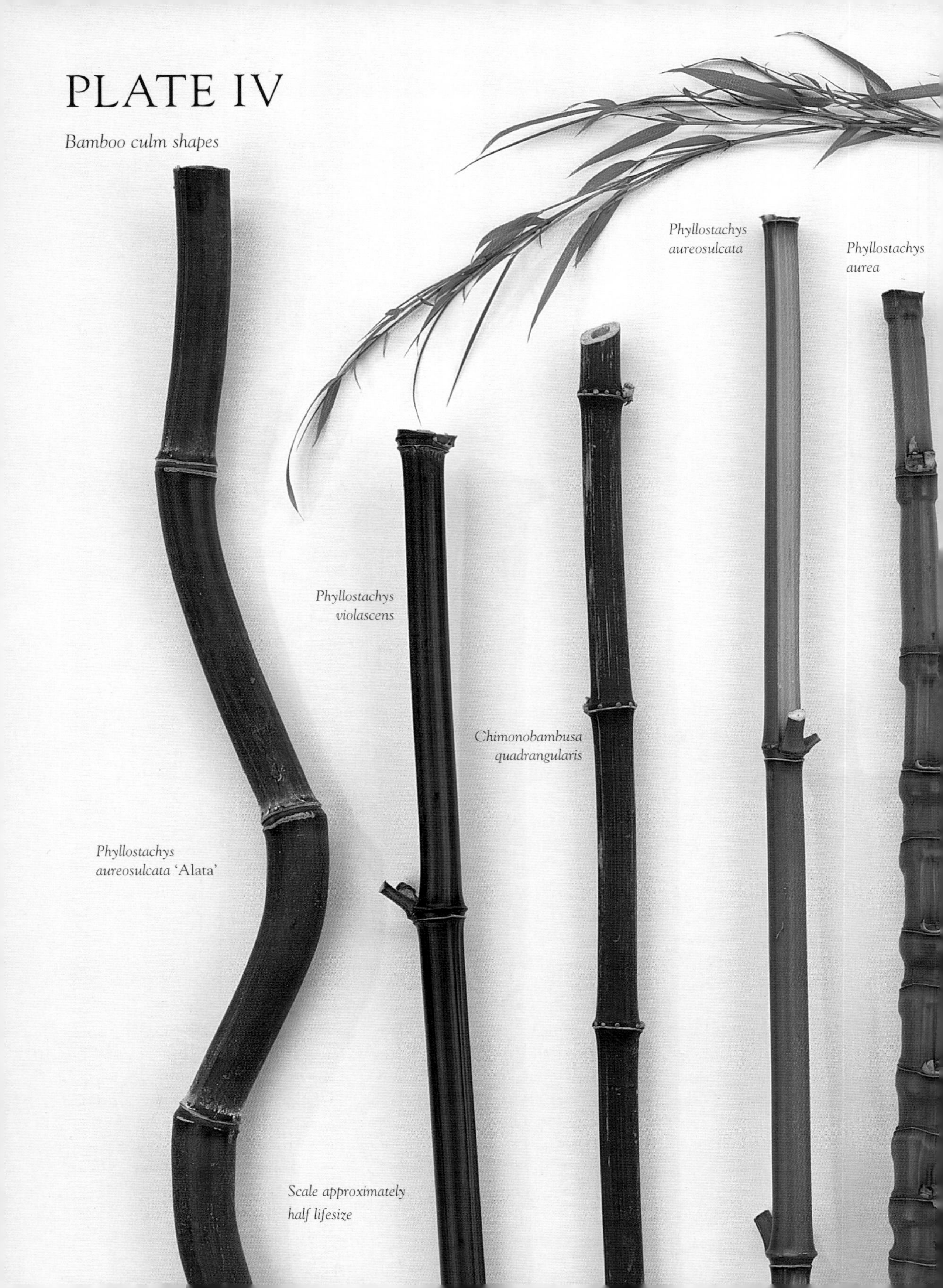

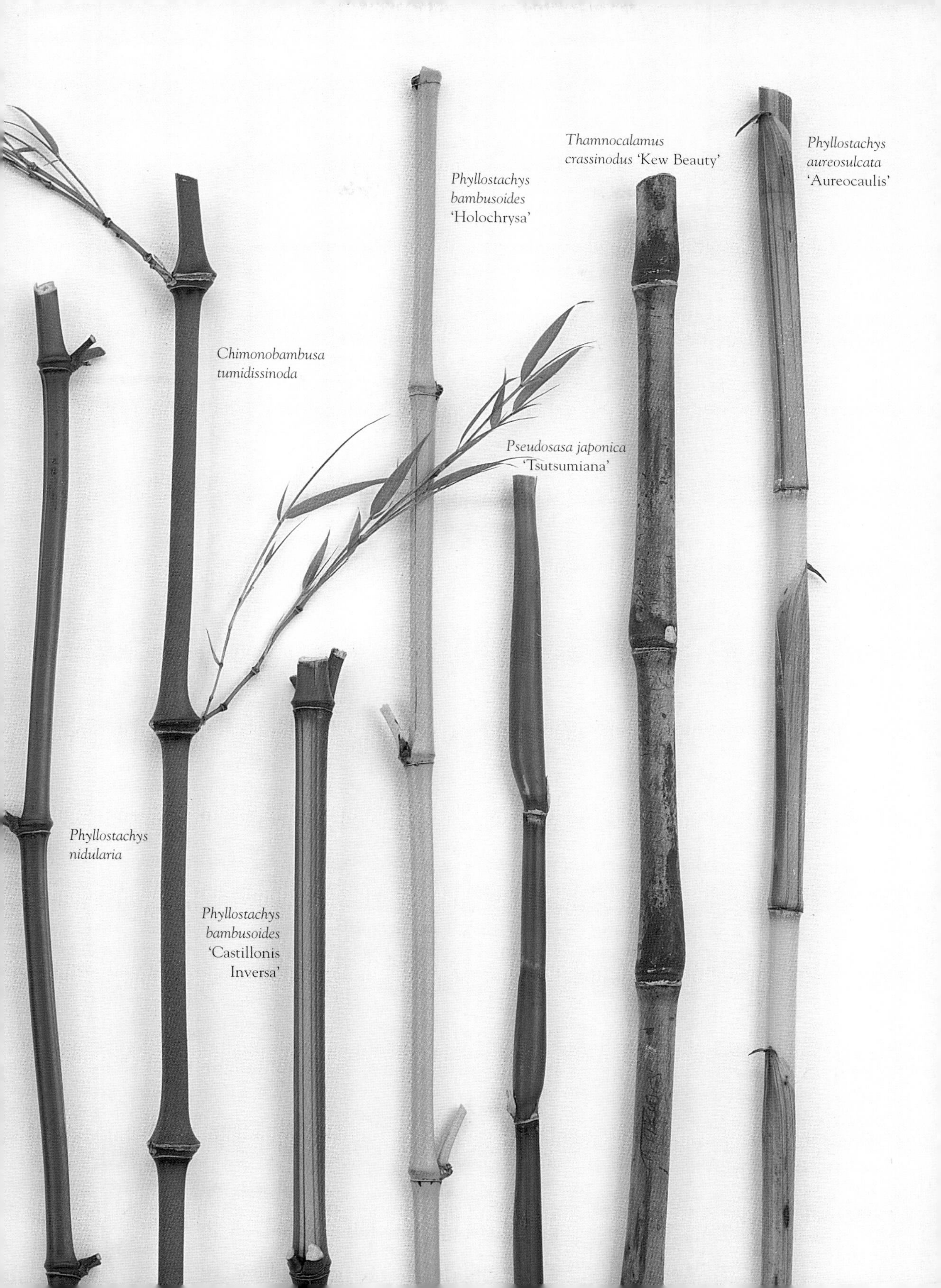

Thamnocalamus crassinodus 'Kew Beauty'
Phyllostachys aureosulcata 'Aureocaulis'
Phyllostachys bambusoides 'Holochrysa'
Chimonobambusa tumidissinoda
Pseudosasa japonica 'Tsutsumiana'
Phyllostachys nidularia
Phyllostachys bambusoides 'Castillonis Inversa'

the outer edge clear the soil away at a low angle. This will be where you drag the plant out of the hole so it should be as gradual as possible. Undercut the root ball for at least two-thirds of the plant width. Usually, if the ground is moist, the fine bamboo roots will hold the soil mass in a good root ball. If this is not so and the soil starts to fall away, the undercutting should be done a bit at a time, and the exposed sections progressively held together with rope and plastic bags. Cut down the two sides and remove or loosen soil as required to free the root ball. Next, lay boards and sacking or bags well under the root mass and then dig down the remaining side. With some pulling on the top growth the complete plant should fall on its side before much soil is removed. If the plant is to be cut into sections this can now be done much more easily, and with much less damage, by working from underneath the root ball with a spade, or by cutting with an old saw. Finally, wrap the root ball with the sacking or bags that are now trapped beneath the root and tie securely to hold it all together during transport.

GROWTH PROBLEMS

Bamboos are remarkably free from parasites and problems if they are growing well, and can easily adapt to less favourable growing conditions. Provided the selected species can adapt to your climate, and as this is not usually problematic, you are unlikely to encounter anything significant. You will rarely need to refer to this section relating to problems and hardly ever need to take any serious actions. Plants in pots need more attention and because of this have their own problems (p.75). If the potted plant is eventually to be planted out in the garden, then this should be done without delay early in the year. The only exception is for plants in very cold regions where bamboos should be as large and woody as possible before exposing them to the elements.

PREVENTING OVERSPREADING

There will be occasions, particularly in warmer regions, where some form of barrier will be needed to prevent a vigorous plant with leptomorph rhizomes spreading beyond its alloted space. In Kew Royal Botanic Garden, sturdy flexible plastic barriers are used to enclose each species. Narrow paths between provide a service zone so that the plants can be tended without damaging new growth and give children visitors 'jungle' paths to explore. These barriers can be purchased from larger bamboo retailers. They should be installed vertically, projecting about 8cm (3in) above the ground and the joints should be bolted together according to the supplier's instructions to prevent the strong rhizomes searching out points of weakness or leaping over the top. This is the ultimate plant control, however, for it is very expensive, difficult to install, and it still does not completely eliminate the need for future care. In a botanic garden, where many species have to be placed close together and nothing can be left out because it is 'unsuitable', this expense is fully justified. In a garden setting it should only be used as a last resort, and then only with plants that are very precious to you. Alternatives, such as large-diameter pipe sections, old metal ducting, reject water tanks or even old rigid plastic garden pools, can be considered but these are usually relatively small in size. An artificial pool made from a flexible liner is no barrier: rhizomes will go straight through even the toughest.

Probably the best solution is to use the growth habits of the bamboos themselves to limit their spread. If unimpeded, the root systems of most mature leptomorph bamboo species grow at or just below the soil surface (a notable exception is *Chusquea gigantea*, syn. *Chusquea breviglumis*, which in most situations has a rhizome system up to 50cm (20in) below ground). When they reach an obstacle, they will go either over or under it, but will return to the same position in the soil on the other side. When surface rhizomes reach a sloping or vertical edge, such as a cutting or the bank of a stream, they continue horizontally for a short distance before they correct themselves and reorientate to follow the new soil slope. So by cutting a sunken path, 30cm (1ft) deep, across the route of an expanding bamboo, you will be able to see all radiating rhizomes when they break the surface and it will be necessary only to cut these with a sharp spade once a year in the winter. You will also have a good supply for rhizome cuttings as a bonus. If a sunken path cannot be worked into your design you could fill it with bark chips, wood shavings or garden refuse, but you will then have to clear it each year before trimming the rhizomes. *Nikon Chiku-Fu* (*The Cultivation of Bamboos in Japan*, p.21) reveals that *Sasa* and *Pleioblastus pygmaeus*, those most persistent of all spreading bamboos, were traditionally controlled with a seaweed-filled trench as most bamboos are damaged by salt. Elsewhere in the book it is stated that 'Seaweed, fish-washings, and kitchen salt do not suit bamboo' and that 'the whole plantation will die off if washings of arame (a seaweed) or buckwheat husks are applied.' According to the author, if *Sasa palmata* is controlled by burning, a common method of removal, it comes back strongly in subsequent years.

The trench system can be used against a boundary with your neighbour, or alternatively plastic-coated metal sheeting of the type supplied for roofing can be fixed below ground to the fencing posts. If you do not do one of these, in a few years you will not only not be speaking to each other, but you will also have the anguish of watching him digging up plants that look better than those on your side of the fence! If you put in a rigid barrier or a trench system, it is essential that you inspect it thoroughly every year. It is not usually necessary to control all sides of a plant as freedom can sometimes be allowed to the rear, or conditions, such as a stream, dense woodland or poor soil, may prevent further expansion on some sides.

It must be emphasized that, for the average gardener, these systems should only be used for special plants and under particular circumstances. It is far better to select bamboos with growth patterns that will not cause you any problems. Also, if you are restricting a plant's natural spread, you should remember to carry out a propagation programme to bring in new vigorous divisions when the captive becomes senile.

Pleioblastus pygmaeus (old form) is a running species and, being not as rampant as most, makes very good ground cover.

Removing invasive bamboos

One of the questions most frequently asked about bamboos is how to remove the rampant legacy of an unwise selection. At any talk on the subject it is always the first enquiry at question time, and is usually followed by 'How can you stop a bamboo flowering?' (p.56). These negative subjects can be very daunting and you need the qualities of a politician to avoid everyone going home thinking that bamboos are potential disasters. Matters are made worse because there are no easy and quick solutions to either question to enable you to regain the initiative. They are very real problems, however, and both need serious consideration.

By the time most people have decided that a bamboo needs removing, it is usually an advanced problem. If the offending plant is a sasa or a pleioblastus consisting of the original compact clump with runners radiating outwards in all directions it is not insurmountable. Dig out the centre clump as described under transplanting (p.51), then loosen the soil around each remaining rhizome radiating outwards from the hole and pull the exposed end upwards. The rest of the rhizome should follow, getting easier at its outer reaches. Hopefully, you will not disturb the surrounding plantings. Phyllostachys and other larger plants can be handled similarly but you will need some assistance, particularly in regions where they reach their maximum height.

It is a completely different matter if the plant has been allowed to continue invading the garden to the point where it covers a large area and has no obvious centre. If *Sasa palmata*, *Pleioblastus humilis*, *Sasaella ramosa* or any of the chimonobambusas have been allowed to cover large areas, it is immediately obvious that the problem is serious and that the only way to act is to employ any means possible, even if the use of chemicals is alien to you. A number of weapons is available and, as each situation will have its own restrictions, a combination of these should be tried.

Any valuable plants within the area should be rescued before work starts if at all possible. A total weedkiller such as sodium chlorate can be used unless there are tree roots in the area or quick replanting is envisaged, in which case several doses of a strong herbicide spray will be more appropriate. The bamboo's aversion to salt (p.55) can sometimes be exploited successfully but this is only slightly less damaging to other plants than sodium chlorate. It has only minor advantages, therefore, but I find it useful for quick spot applications to kill persistent pieces, regrowth or outer extensions. It is very unlikely that one session of any treatment will be sufficient to eradicate a well-entrenched bamboo, and repeated applications next year and constant inspection for new growth over several years is an essential follow up.

All invasive bamboos are leptomorph species so the food reserve in the extensive rhizome system ensures active but smaller regrowth if the culms are regularly cut off. Repeated cutting will seriously weaken the plant but other treatment is required to prevent it rapidly re-establishing itself. This weakening makes bamboos much more sensitive to other controls, however, so a combination of systems has great advantages. Not only are the plants more sensitive but the soft new growths absorb sprays much more effectively.

If cutting large areas of tough culms is a problem, you might consider burning or using the type of flame gun that is sold for weed control. Bamboos are very sensitive to flames because they have no protective bark or large quantities of surface sap, and flame guns are very effective on smaller culms. However, the end result is a mess that has to be cleared away.

A technique that has been reported as successful for larger species, such as phyllostachys, involves cutting all culms at ground level, and then cutting all regrowth culms horizontally and cleanly at waist height and between nodes. The 'cups' formed by this treatment are filled with a herbicide solution before they have had a chance to harden.

Most of these problems result from earlier times when the potential growth pattern of imported species was unknown, or when there were few people with experience to consult. The best modern solution is to avoid the problem by selecting species that are not going to outgrow their welcome in your climatic region. If you are unsure always take the advice of, and buy from, one of the many specialist bamboo nurseries now in existence, rather than opting for the mass-produced plants from a garden centre. The tempting cheap plants sold in quantity from the latter are usually so because they grow very fast.

Preventing flowering

If a problem plant starts to flower this is likely to be seen as a welcome development – any disturbance, spray, or additional stress at this time is likely to be the *coup de grâce*. If you just wish to remove the now-unsightly specimen (as this phase may last several years) this is easily done by cutting it to ground level.

If the flowering plant is one that you have cherished and wish to keep, the onset of complete flowering is a sad occasion. Strong-growing leptomorph species usually recover in time but weak specimens, new plantings and pachymorph species are at great risk. All that can be done is to protect the plant from any additional stress and feed well so that it is best able to cope with this phase. Cutting off the offending culms is often recommended but this also removes the food reserves that are supporting the process and also what little chlorophyll remains now that all the leaves have gone. If the flowering is partial, removing the culms is of little consequence and improves the appearance, but if the flowering is complete it cannot help the plant. There is no known method of foreshortening or preventing flowering at the moment and the only additional actions that can be taken are to plant a possible replacement species in front of the offending plant and to collect seeds.

Soil erosion

Although it is necessary to be very careful when introducing the more vigorous leptomorph species into the average garden some can perform a very useful function in landscape applications. The web of extremely strong connecting rhizomes produces an almost indestructible carpet just under the surface of the soil. This is ideal for stabilizing newly exposed areas of soil and even sites at the top of cliffs, provided the ground is not too dry. Most bamboos are not seriously harmed by short periods of flooding. Indeed, some species such as *Pseudosasa japonica* or *Sasa palmata* are ideal for stabilizing river banks and are used in reclamation schemes where there is land that is liable to seasonal flooding. In countries where extensive logging takes place, bamboos may perform a useful function in preventing the loss of valuable topsoil.

Pleioblastus pygmaeus (left) and *Pseudosasa owatarii*, here in pots, are dwarf clones that make useful garden plants and both can be clipped to form a very low mound.

Aborted culms

Concern is often expressed over the aborting of new growth mainly on established but immature plants, either at ground level or as the culm is expanding. This is a reflection of changes in growing conditions or an imbalance between the above ground and the subterranean parts. To some extent this is to be expected with any bamboo, particularly young specimens, unless conditions are constant and perfect. It must be remembered that the energy for this year's growth is supplied by last year's growing conditions. If the expanding culms run out of a supply of nutrients or fluids halfway through the expansion stage, they will stop growing and any buds at ground level will fail to develop. Culms and rhizomes also react to different stimuli. Roots, logically, seem to be stimulated mainly by soil temperature and phosphorous fertilizers, while culms respond to good growing conditions and nitrogen. Therefore, a young plant can have a juvenile root system but produce energetic culms, the result being a compact plant with substantial culms. Conversely, a running (leptomorph) rhizome that is mature in character can be supporting small-diameter floppy culms. These variables can cause aborting culms on new plantings.

Leptomorph bamboos often develop many more small buds than are actually needed to form future culms or additional rhizomes (this may be true of other bamboos, too) and most never develop beyond the bud stage. However, excessive fertilization seems to stimulate more of these into activity than the plant can support. The resultant growth is soft and very prone to significant amounts of aborting or other growth disorders. Aborting can also be caused by any stress to the plants during the culm expansion stage, such as high winds or drought.

Site problems

If the plant is not suited to your climate, or to the position that you have selected for it, it will usually tell you in obvious ways. If it is too hot, too dry, or the light is too strong, the leaves will curl within minutes. The cause can be quickly established by watching to see whether the curl is temporary or permanent, and if it can be overcome by watering well. If watering cures the problem, dry soil is the cause, and this is not always immediately obvious from the soil surface condition, particularly with plants in small pots. Permanent leaf curl can only be cured by moving to a cooler well-drained or more shaded spot.

Lack of light

Sparse but healthy leaves and lack of growth is usually due to insufficient light. Similar symptoms accompanying a sick-looking plant is most often caused by bad soil or poor root condition, more usually both, and more often than not this is caused by poor drainage. In northern latitudes, some more equatorial plants, such as *Chusquea coronalis*, can suffer from lack of light, even in full sun. These plants may just fail to grow. More seriously they may be unable to achieve the necessary strength to withstand temperatures even a few degrees above freezing, and will therefore die during the first winter. The seemingly most unlikely species can also suffer from a lack of light and fail to achieve their normal luxuriance. The high-altitude homes of some mountain species have a much greater light intensity than many of our more northerly gardens. This is partly because of their latitude, but also because the light intensity is greater due to the thin atmosphere. Although we read about them naturally growing in open woodland swirled in mist, we should not necessarily conclude that a shady site is best. This is often the case only in hotter areas nearer the equator.

Lack of water

Lack of water at the roots results in lack of growth during the summer period. Most species have a period of rhizome growth where nothing seems to happen above the soil level. The time of year varies with the species, but overall a steady growth should be achieved. Established plants fall dormant during short drought periods, but new plantings will not prosper without a steady supply of moisture at the roots. Most temperate species come from regions with much heavier rainfall in the summer than they will experience in cultivation and so for maximum growth additional water is advisable.

Low temperatures

Some species with running rhizomes, such as the *Phyllostachys*, may not develop their characteristic energetic root system, particularly in cooler regions. Often this is an advantage in the garden, except where a grove effect is desired. Assuming that the plant is otherwise healthy and that there are adequate water and nutrients in the soil then the rhizome activity is being curtailed by low soil temperatures. This will be improved by allowing the sun to reach the roots and by dark mulches. A sunny, south-facing aspect is also an advantage, together with protection from the wind, but these remedies often involve relocating the plant. It has been written that running rhizomes always expand in the same direction, but, in fact, rhizomes are more active on the side of the plant that has the better growing conditions. They usually grow into the warmest soil, which is more or less on the side that the sun shines most, but they can seek out the wettest, or most fertile soil, or expand all around if conditions are uniform.

Dry air

Plants in locations that experience dry air conditions, particularly with high winds, can exhibit temporary leaf curl, as with too much light or too little water. In these areas a bit of shade, wind protection and ample water to the roots is more important than ever. There is a hidden danger, however, with those species that prefer an oceanic climate. Dry air is usually associated with continental-type climates that also have cold winters. This combination can be lethal to the plant. The dry air has a debilitating effect on the plant, weakening it so that it is killed by the winter. This is true of the chusqueas, and presumably some other mountain species, which are classed as very hardy and wind tolerant in areas with a coastal influence, but are often much less so inland.

This should be remembered particularly when planting a hedge as the plant will be very exposed in such situations and the quoted minimum temperatures should be adjusted accordingly. The evergreen qualities of a bamboo are one of its main assets and this can be ruined by the wind or by the cold. Also variegated culms can be stained brown by the cold, and all of these factors reduce the temperature range over which ornamental species can be used.

PESTS

Problems with insect attack were, until relatively recently, rare and not a problem. Slugs and snails will sometimes eat the soft parts of the expanding culm where they are exposed by the tough culm sheath. The damage is not always recognised as such, as the resulting hole, which can cause distortion of the culm, is considerably higher up the culm by the time it is usually seen. Aphids are the other common problem but are rare on most species of bamboo. They have a particular liking for *Phyllostachys bambusiodes* and its forms, and on this species it can cause sufficient damage to warrant treatment, particularly in the more reluctant-growing forms. Scale insects seem to like some other species of bamboo, and this can be a problem on plants growing in the greenhouse. It is rarely more than unsightly on garden plants.

Over the 1990s the bamboo mite (*Schizotetranychus celarius*) has rapidly spread from its native Japan, across North America and then into most of Europe. Although this does little serious damage, it is very unsightly and a cause of great concern to the gardener, and even more serious to plant nurseries. The first indication of trouble is the appearance of unusual variegations on the older leaves of the plant in 'morse code'-type bands following the leaf veins and about 3mm (⅛in) wide. An examination of the underside of the leaf shows 'nests' of the mites protected in webbed capsules under each yellow streak. They are more prevalent in areas with low air humidity, but in all areas great care should be taken over the purchase of new plants and a system of quarantine introduced. These precautions should be taken even for apparently clean plants, as it is usually only in the leaf's second year that obvious damage can be seen. A combination of removing old culms and leaves and spraying repeatedly with a systemic insecticide seems to be the best way of attacking an infestation in small plants. Very vigorous bamboos, such as sasas, could be treated by removing all the top growth, but this would not be wise for weak, slow-growing or compact plants.

Animals can cause more serious damage. Attacks are infrequent, but when an animal has acquired a taste for the succulent shoots and leaves, it can cause considerable damage by repeatedly removing any new growth. Squirrels can be a nightmare as they consistently remove the tops of the new shoots as soon as they appear above ground. Protecting the new growths is not a long-term solution as the squirrel or other animal will simply remove the tops as soon as the culms rise above any netting. The recently introduced electronic water spray systems that are activated by an approaching animal would seem to have potential, if the devices are reliable. The only permanent solutions are usually drastic (e.g. trapping) for it is almost impossible to protect large plants from persistent attack.

Animals that dig at the roots of bamboos are usually after the worms or insect life that collect there because of heavy manuring, or they are attracted by fertilizer smells. Stop using animal manure, bonemeal or other organic material, and either do not fertilize for a while or use a chemical fertilizer. A layer of chicken netting hidden just under the soil is very effective but can be a problem if you wish to divide the plant in future years when it has expanded and grown through the mesh.

4
PROPAGATION

Those new to bamboos always find it an anomaly that, despite the fact that they are known to be the fastest-growing plants, they are among the more expensive to buy, and are not available in huge quantities. There is a good reason for this, however. The more desirable garden bamboos are, naturally, those that grow slowly at the root and have rapid vertical growth. The slow rate of root growth means that the best one can hope for is that the plant will approximately double its size every year. Therefore, for every plant in a nursery, one can expect to get only one similar-sized division per year, not allowing for any propagation failures or for the plant to increase in size. Micropropagation is not yet suitable for many bamboos because of difficulties in isolating the meristem (growing point) and because some of the more desirable plants are variegated. With improved techniques, perhaps it will not be long before these problems are overcome. In the meantime, traditional small-quantity propagation is the main way to supply the demand.

DIVISION

Perhaps the easiest and most reliable way to produce an extra plant or two is to take divisions. If they are to be taken from mature plants growing in the garden the previous sentence is a huge over-simplification: I keep sharpened spades, a large axe, heavy hammers, strong loppers and special cutting tools for this purpose! And eye protection is essential. By contrast, divisions from pot-grown plants is relatively straightforward.

When to divide

Timing is important, for although big divisions can be taken most times of the year if the conditions are moist, except for very hot or very cold periods, smaller pieces are not nearly so accommodating. In most regions early spring, just before the rhizome buds are beginning to expand, is undoubtedly the most favourable. This usually coincides with the very first signs of growth to the leaf and branch buds. Once it has been divided the production of new shoots will be delayed until the plant is established, and the size of the buds, and so the potential culms, will be adjusted to suit the new, less favourable conditions. When spring is well underway, a time when bamboos are normally sending out the new culms, the plant will be fully adjusted to its new surroundings and ready to go. If divisions are taken at this later stage, the bamboo will be all prepared with much larger buds ready to expand into full-sized culms, when along comes the keen gardener and cuts off most of its stored energy supply: the new shoots will usually abort and will not be replaced for many weeks.

In areas with mild winters and long autumns, early autumn is also a very good time to make divisions, for the plant will have the necessary humid air conditions and warmth at the root to enable it to become established before winter sets in. When spring arrives, it will be ready to respond unchecked to the improving weather. If the autumn is short or the division is not strong enough before winter arrives, however, it will be difficult to stop the roots rotting over winter. Allow at least four weeks from making the division to the onset of winter. If this is not possible, then delay dividing until spring. Along with all other potted bamboo plants, potted divisions should not be left unprotected outside overwinter, even in relatively mild areas.

If you garden in an area with very short springs and autumns, be guided by the growth of the plant. In the

PLATE V

A bamboo growing from seed (Fargesia murieliae)

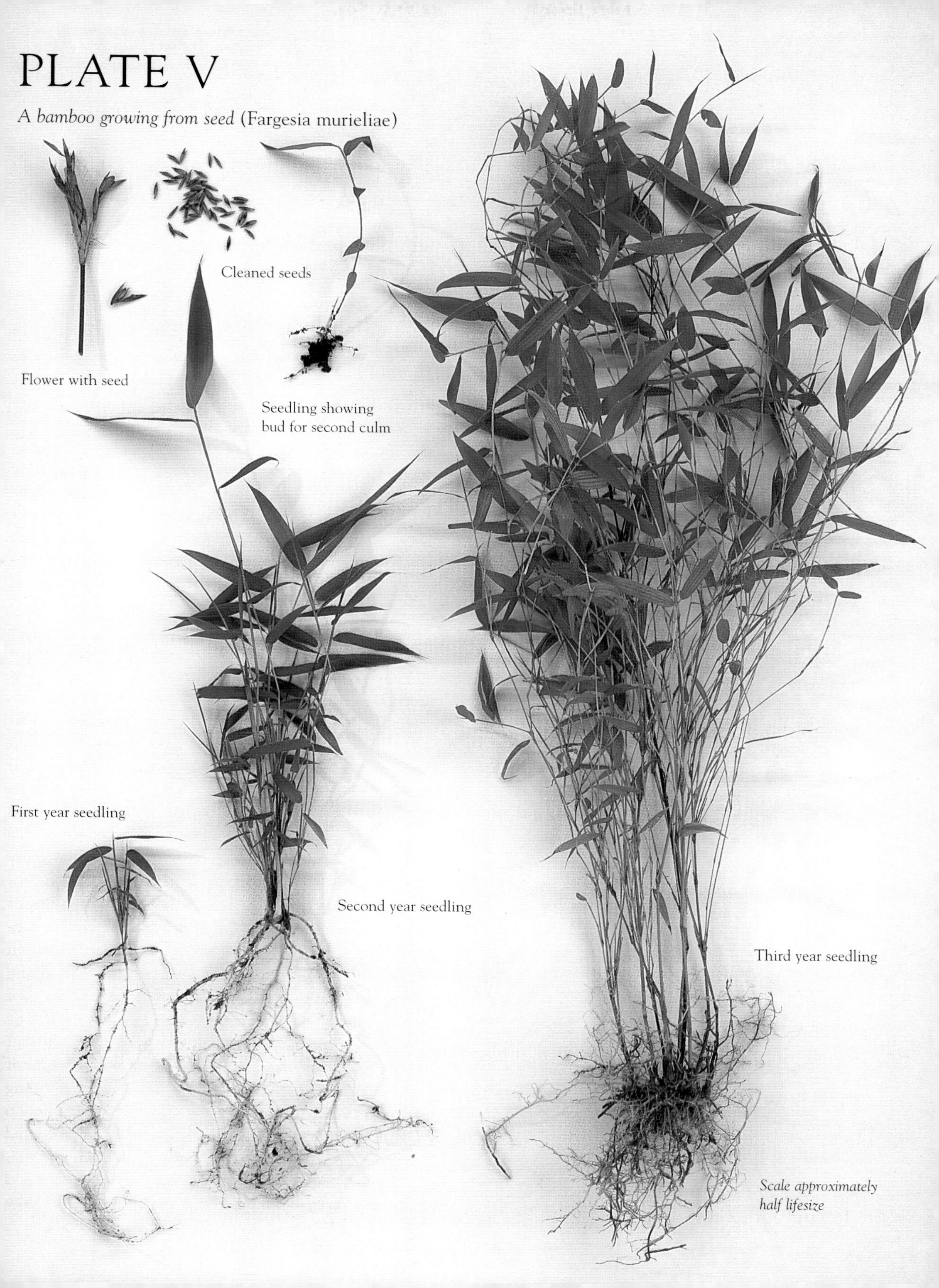

spring make divisions at the very first signs of growth and, if the weather outdoors is not favourable, create suitable conditions artificially in a greenhouse or nursery area. Divisions can also be made in early autumn and the new plant kept frost-free over the winter.

Dividing plants in the garden

In mature garden plants the root ball consists of almost solid rhizome, on which normal tools make no impression, so usually it is best to inspect the perimeter of a clump. First look for a few culms that are close together and appear to be from the same piece of rhizome, but which are only connected to the main plant by a single rhizome. Before cutting, remove most of the soil from around the group and check that when one culm is moved the adjacent culms also respond. If they do not, they are probably on different rhizomes and are, therefore, not suitable. Try to select a group of at least three culms with a good complement of leaves and not too old. Culms without leaves generally do not have roots, and old culms have often lost the ability to produce new growth. As mentioned earlier (p.37) the cells of old culms and rhizomes have probably not reproduced since the culm was formed several years earlier.

Once the interconnecting rhizome is severed with a sharp spade or loppers, the lifting of the division is fairly simple because of the compact roots. Try to take as much soil as possible with the roots. Only remove as much of the top growth as is necessary to ease transport or to balance a small root system if this is found upon lifting. The starches stored in the culms are the main source of energy available until the roots become re-established and any unnecessary trimming could hold back the size of the new culms and the speed of the bamboo's establishment in its new surroundings.

Larger divisions of more than three culms can be reliably taken from established clumps during autumn and spring, provided the ground is moist and there is no

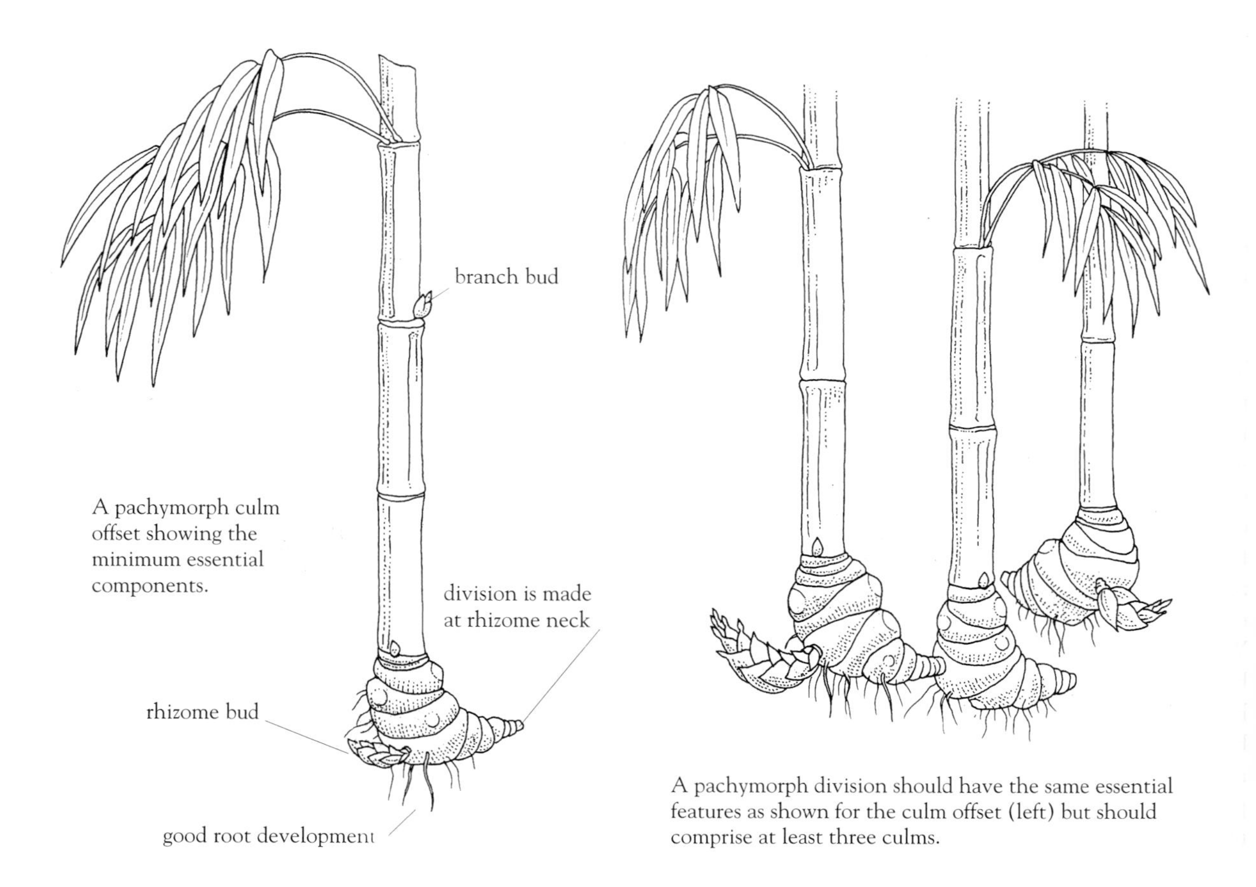

A pachymorph culm offset showing the minimum essential components.

A pachymorph division should have the same essential features as shown for the culm offset (left) but should comprise at least three culms.

vulnerable new growth to be seen. Study the clump for any isolated sections that have formed naturally, and accentuate this isolation by removing any culms that are growing along the division line. Pull or tie the two sections apart and cut the rhizome. Sharpened spades, axes or wide masonry chisels are normally required to do this. Throwing a heavy spade at the division line is a technique often used but this is very hit-and-miss, particularly as you get more tired, and considerable damage to the prospective division can result. Hitting a static chisel or axe-head with a heavy hammer is more accurate.

Tools designed specifically for the purpose of making bamboo divisions can be made by a local garage or blacksmith with good welding facilities. They are well worth the expense if you do this type of work frequently as the risk of injury is reduced and the divisions produced are much less damaged than when obtained by other means. For large divisions I use a masonry chisel which has had its shank welded to a flattened steel tube about 6cm (2½in) diameter and 1.5m (5ft) long. A loose heavy solid metal bar slides inside the tube and acts as a hammer on the chisel handle within. This device is then located on the rhizome to be severed and hammered repeatedly on the same spot until it is cut through. The operation needs repeating several times along the division line before the division can be levered free. The chisel is sharpened repeatedly as it blunts quickly on the rhizome, soil and stones.

Dividing potted plants

To divide a pot-grown plant, first remove it from the pot and, if the root system is healthy, choose a site to make a cut that will do the least damage. Using a hose wash the soil off from around the site to reveal as much of the rhizomes as necessary. If the root ball is open, dividing the plant is then simply a matter of using secateurs. If the plant is filling the pot with a mass of rhizomes and roots, and there are many culms, it is often easier just to cut the mass into sections with an old saw. When deciding on the size of the division, a much-used but very accurate saying should be remembered: one healthy division that can be divided again next year is better than two dead ones (or two that take years to recover). Repeated division of this sort encourages the plant to develop smaller rhizomes and culms which can make future divisions progressively easier.

Culm offsets

A refinement of division is culm offset cutting. This is simply a small division comprising a single, shortened culm with fresh rhizome buds and roots. As with divisions, the culms used should not be less than a year old but should look young and healthy. Culm offset divisions require the culm to be cut back to just above a convenient node to balance the rate of transpiration through the leaves with the small root system. Consequently any new growth will be much smaller than the parent and the cutting will take a few years to become a good plant. Try to find a culm with branches and leaves low down and then cut above a node with branches, or at least above a node that shows viable branch buds. It is then easy for the plant to form new leaves as it recovers. If there are no branch buds on the cutting, the new plant will have to find energy reserves to put out a complete new culm, branches and leaves before it can transpire effectively. Usually it will be unable to do this.

As when taking divisions, inspect the mature clump around the circumference for a suitable isolated culm. It should have plenty of leaves and should be fresh and colourful. Frequently the outer single culms are this year's growth and have an undeveloped root system and few leaves – particularly so with the genus *Yushania* – and these should be rejected. It is often necessary to delve deeper into the clump to find good material. The single connecting rhizome is usually not difficult to sever with a sharpened spade but if the special tool is available, it can make things considerably easier.

With pachymorph plants the cutting can be taken at the rhizome neck. This should be found by making the cut as close as possible to the culm that is radially next to the one that is going to be lifted. Leptomorph plants are physically easier to select and divide because of the wider culm spacings, but often more difficult to get a good division. The length of the cut rhizome should be as long as possible: with vigorous species this can be up to 1m (3ft). It has to include at least one and if possible several rhizome buds so that it is able to produce new rhizomes and culms. A common problem with leptomorph species is lack of roots on the section of plant lifted. When the division is exposed for cutting, if it seems to have too little root, retain as much rhizome as possible, but if there is plenty of root, the rhizome can be trimmed to a convenient length. Those species with thin rhizomes can have them coiled around the pot but species with strong rhizomes should be put in a large pot or planted in a sheltered spot in the garden.

Growing on

Small divisions and culm offset cuttings should be kept in a stress-free environment until new growth is seen. A cool, humid greenhouse is ideal, but if this is not available, a shaded wind-free spot in the garden is good if your air humidity is high.

During this stage leaf curl often occurs, and is to be expected to some extent as the plant recovers from the shock. Watch the division carefully for the first 24 hours and be ready to take action to prevent all the leaves dying should the leaf curl become severe. The best course is to trim off some of the leaves, in proportion to the amount of stress, leaving branches and twigs so that the leaves can be replaced easily by the plant as it recovers. The first few hours in the life of a division are most important and, for example, a journey in a hot car has killed many an otherwise good potential plant. During transport or between lifting and potting, keep all divisions wrapped in plastic and well watered.

If the division is good and taken at the right time of the year, it is likely to recover quickly, and can be potted into a proprietary potting compost with a coarse structure or with added grit to assist drainage. In the case of small or poor divisions, however, the traces of fertilizer contained in these composts are liable to damage the roots before the plant has recovered sufficiently to absorb them. Pot this type of division temporarily into an inert compost such as coarse peat. Water them liberally every day until they have recovered and, when new growth is observed, introduce a liquid feed until it is time to pot them into normal compost.

RHIZOME CUTTINGS

Leptomorph species can be propagated by rhizome cuttings, which is a very good method of propagating larger quantities of plants. Cuttings of young rhizomes, no more than two years old, should be taken in early spring in sections 15–30cm (6–12in) long and containing at least two dormant buds. Place them, sloping, in a tray of compost with the uppermost bud just at the compost surface. Within a few weeks new culms should have developed from the buds if warm conditions are provided. Although as much heat as possible is needed at this stage, in cool climates there is no need for light until growth is seen. The tray can be placed in a plastic bag and kept in a garage, warm shed or even the house. The size of the initial culm is limited by the food reserves in the rhizome section and the new culms will be about 1.5m (2ft) high and of small diameter, but given consistently good growing conditions should have made reasonable plants within two years. Rhizome cuttings taken from species with small-diameter culms, such as sasas, can be cut into smaller sections, each with a couple of viable buds. They will be very slow to make good-sized specimens, but this system is ideal for producing a large quantity of clean plants.

For species that enjoy warm summers, such as the *Sinobambusa* or *Phyllostachys*, this method of propagation is usually only successful in the warmer temperate areas. Not only are these species reluctant to co-operate in cool conditions but also vigorous rhizome sections are not always available.

In cool areas it is easier and more reliable to make rhizome cuttings on the plant. In early spring expose a rhizome by sweeping away the top layer of loose soil, for they are usually only just below the surface. Cut the rhizome into suitable lengths without lifting it, and re-cover it with soil. Water and feed during the summer and by the end of the year the cuttings will have formed small plants that can be lifted the next spring. This method is by far the best for propagating some of the *Pleioblastus*- and *Sasa*-type species that send out long runners with very few roots attached. Not only are long rhizome divisions of this material difficult to accommodate if potted conventionally, but they are also slow to recover and slow to make good plants.

SEED

When seed is available, it is a very good way of raising a reasonable quantity of clean new plants that, hopefully, will not normally flower during the lifetime of the gardener. Seedlings are often very variable so there is also the chance that something special will grow, or at the very least we will obtain plants that are more suited to our climate. The seed should be sown as soon as it is collected (p.44) or obtained as it rapidly loses its vitality. Use normal seed compost with up to 50 percent added coarse grit or vermiculite to give good drainage properties. It will usually germinate within a few days, and keeping the seed tray in the dark seems to

PLATE VI

Bamboo sheaths – rhizome, culm and leaf

Scale approximately half lifesize

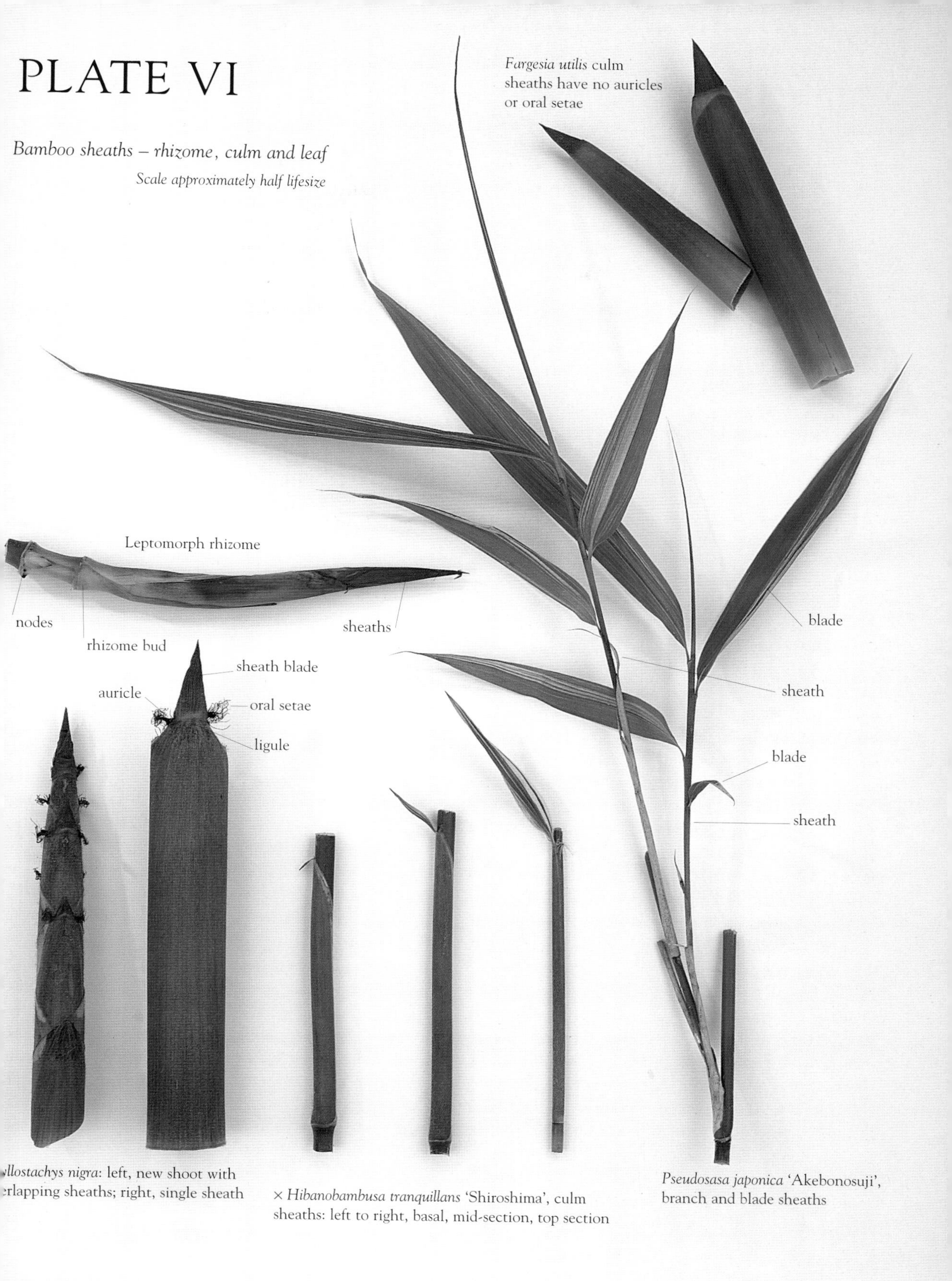

Fargesia utilis culm sheaths have no auricles or oral setae

llostachys nigra: left, new shoot with rlapping sheaths; right, single sheath

× *Hibanobambusa tranquillans* 'Shiroshima', culm sheaths: left to right, basal, mid-section, top section

Pseudosasa japonica 'Akebonosuji', branch and blade sheaths

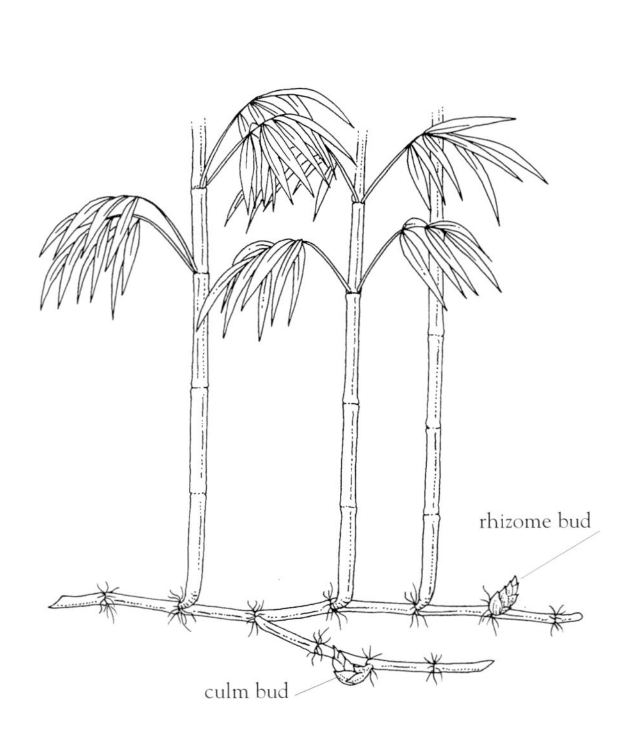

A leptomorph division should comprise at least three culms and have the essential features shown for the offset (right).

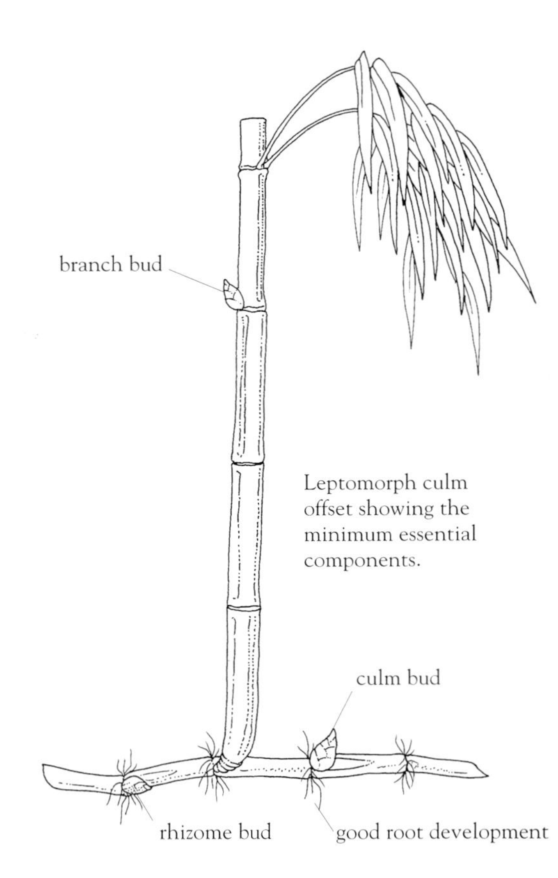

Leptomorph culm offset showing the minimum essential components.

be of no benefit. It can, however, take many weeks for seedlings of some species to appear, or even a year or more with a few types. This can mean that you have to nurture tiny seedlings through the winter with all its associated dangers. To avoid losses, I normally place the seed tray in a small propagator on a window shelf of an unheated, south-facing room in the house, where the seedlings can continue growing steadily and be given closer attention.

Sometimes the seed of species from the mountain regions will germinate better after a period of cold. If this is a known feature, place the seed in the refrigerator for a week or two before sowing. It is advisable to store seed in a refrigerator anyway if sowing is delayed more than a few days in order to retain viability for as long as possible. Alternatively, if seed is collected late in the year, sow as normal and put the seed trays in a cool greenhouse overwinter, then good germination should be obtained next spring, even with those species needing a cold spell.

Once you have achieved seed germination, damping off of the seedlings is the main problem you are likely to encounter so use the normal chemical precautions, such as Cheshunt Compound; alternatively, keep the seedlings slightly on the dry side until they reach a reasonable size. The only other problem that may be encountered is if the seedling fails to get beyond the first leaf stage. Usually this is due to suspect seed that has been stored for too long or has been collected before it has reached maturity. It can also be because the right conditions have not been provided to enable the young plant to prosper after the nutrients in the seed have been used. To avoid the latter take care that the compost does not stagnate by incorporating ample grit or vermiculite in the mix and watering with care. Faster-growing seedlings usually need potting on frequently so soil problems are not then a problem. Some seedlings resent high soil and air temperatures, particularly the pachymorph genera, and a shady spot in a well-ventilated greenhouse or even in the garden gives good results. When the young plants comprise many very fine culms, they are ideal for division if additional plants are required.

BRANCH BUDDING

Some tropical species are readily propagated from the branch buds. As the culms age, the number of branches increase and small rhizome-like structures are formed at their bases, sometimes even accompanied by aerial roots. As stated earlier (p.31) there is little botanic difference between the various parts of a bamboo, so if sections of the culm are cut and placed in warm soil, these rudimentary structures will sometimes produce new rhizomes, roots and culms.

It is important to distinguish between these rhizome-like structures and the aerial roots that are occasionally seen on culms, and even on the lower branches, of some leptomorph species. Normally, these aerial roots are not found in conjunction with rhizome-like structures and, although some success has been had with rooting the nodes, they are not able to form new culms and do not prosper. Perhaps these roots without rhizomes are akin to the similar roots without rhizomes on the base of leptomorph culms.

There is room for experimentation with branch budding on some temperate species that are closely related to tropical species. I have had success with *Himalayacalamus falconeri* 'Damarapa' under mist on a heated bed and a colleague has layered *Himalayacalamus equatus* (not in general cultivation) by fixing a living culm horizontally in a mist bed. *Chusquea gigantea* (syn. *C. breviglumis*) has been observed with these structures and desiccated roots at the nodes, but so far I have had no success with propagating it in this way. As *Chusquea culeou* in all its forms is thought to be closely related, it may be well worth persevering with attempts to propagate it by branch budding: it is not easily propagated by division and hence commands a high price. According to the experts, the trick is to choose the node at the correct age and the right time of the year, which might sound simple until you try.

TIPS FOR SUCCESS

Apart from failures due to incorrect timing, the main problems with all types of propagation are associated with stagnant compost. If the propagule can be seen to be healthy and vigorous, and is likely to continue growing without check, it can be potted on as normal or even planted out into a sheltered place in the garden. Most small pieces, however, will need a period of rest before they regain their vigour. During this period, which can last several weeks, they do not need fertilizer and there is the danger of the compost becoming stagnant. Proprietary composts can be used for propagation, but they must be low in fertilizer and the texture should be opened by the addition of coarse grit, coarse composted bark, large particle peat, perlite or any similar material. Compost with a coarse structure needs only small amounts of additional material, whereas fine composts can have up to a third of roughage incorporated. Fertilizer should only be provided when the plants are growing well and able to utilize it quickly. Seedling compost must be fine in texture, but should be made free-draining by using equal parts of fine compost and perlite, vemiculite or coarse sand.

All young plants will benefit from a period in a greenhouse, if this is available. At first keep them shaded, cool and humid. As they gain strength increase the ventilation and also the light intensity in more northern latitudes. Also of benefit, but far from essential, is a mist unit with a heated base, for the first days of propagation. These facilities enable the most unlikely pieces of plant to be propagated and speed up the production of strong, healthy specimens.

5

BAMBOOS IN THE GARDEN

Every garden is a work of art, whether it is an exhibition display, a grand estate or a tiny back yard with space for little more than a washing line. It is an art form probably more exacting than any other. A painting is a compilation of form and colour on a flat canvas, a sculpture adds a third dimension, architecture has to respect the needs of the everyday world in addition, but a garden has something of all of these and is a living thing, too. A garden is a place to live in, to experience, to touch, smell, hear and feel, to acknowledge the passage of time and look forward to the possibilities of seasons to come. Above all, in this pressurized modern world, it is a place into which to retreat for a few moments and leave the noise and bustle far behind.

The materials used in forming a garden grow, change with the seasons and introduce other elements such as scent, movement and sound. These materials can, if used with skill, elevate the whole range of senses infinitely more than any still life. If a garden is to succeed, the gardener has to learn and respect the needs of every plant, as well as those of the local wildlife. The requirements of its living components (human and plant) dictate the garden's framework, reducing the options and giving the design a direction which is instantly understood by all who see it, even if subconsciously. Every garden should have a grand design: the more humble the plot, the clearer should be the guiding star.

The gardener should select the materials as carefully as an artist chooses and applies every colour. In a restricted space every plant should earn its keep and possess as many qualities as possible – colour, scent, leaf shape and overall form. All the features should contribute. If they do not, you have the wrong plant, or it is in the wrong garden, regardless of its merits. It is clearly wrong to be swept along by the undisputed beauty of the flower of a modern rose, for example, if it is a weak variety or its overall form is not pleasing. Much better to choose something like the tree poppy (*Romneya coulteri*) with its beautiful poppy-like, fragrant flowers, attractive glaucous foliage and elegant form. And both of these would be wrong for a woodland glade.

Plants that are elegant in leaf and form, evergreen, completely hardy and well behaved are very few and far between, reducing the choice of plants that would suit any garden design down to a very few. The majority of bamboos possess all of these properties and, if selected wisely, they can be of great value in all garden settings, except for the arid and permanently windswept regions. Unlike most evergreens, which have thick, glossy leaves or are coniferous, the leaves of bamboos are delicate and rustle in the wind, providing a further dimension.

When growing luxuriously as here, *Yushania anceps* makes one of the most elegant garden plants.

BAMBOO HEDGES

Most gardens need some form of enclosure, either to give privacy or as protection from the wind. Both are essential if the garden is to be enjoyed and lived in as much as possible. Bamboos make excellent screens. Unlike some hedging plants, their roots are not far reaching (avoiding problems with damaged drains and undermined foundations) nor do they produce a desert zone where nothing else can survive. They can be selected to grow as upright as any clipped hedge, although naturally not so uniform and, consequently, not so difficult to place in a less formal setting. Species can be chosen to reach maturity rapidly, probably faster than most hedging plants, and when they reach the desired height, they do not need constant trimming to check a relentless upward growth. Also they do not lose leaves at the lower levels as they age. Most bamboos of reasonably upright habit will make a good barrier and, if space is no problem, even arching species can be used for screening. It is more a question of the density required, with the clumping species giving almost one hundred percent screening, and the phyllostachys and yushanias an open but very effective visual barrier.

It is rare to see bamboos as a clipped hedge or specimen, but they can be trimmed very successfully if necessary. Height is usually the main reason for clipping, although this can be avoided by selecting the right species in the first place. To limit the height, clip back the new culms every year, as soon as they rise above the chosen level. If they are left any longer, they will put energy into unnecessary upward growth when they could be producing low-level leaves and branches. The sides can also be clipped, if desired, and the resulting hedge will be lush, dense and bright green, so long as a mulch and some fertilizer is provided to compensate for the material that has been removed. Hedges also need their old growth to be removed periodically, and to be watered or fertilized, just as any other plant, if they are to remain effective, attractive and able to withstand harsh conditions. Remember, however, that too much fertilizer will produce new growth that will be soft and lush and quickly damaged.

To avoid the disappointment of your nicely maturing hedge bursting into synchronized flowering, I strongly recommend making all windbreaks and visual screens from a mixed planting of several species. Even if the plants eventually recover from their flowering period, it will take them a long time, and in the meanwhile you will have wasted several years in addition to the few that it takes for the plants to reach maturity in the first place. The alternative is to use species that have recently flowered: *Pseudosasa japonica* and *Yushania anceps* are both suitable, and *Fargesia murieliae* will be soon. This may seem to be taking caution to the extreme, but I have twice experienced flowering screens, and I always regret those lost years, when my garden was open to the winter gales for lack of a bit of forward planning.

No bamboo can tolerate dry winds, so the following recommendations are made on the assumption that the site, if very exposed, is open to oceanic winds, or to less stressful conditions inland. It should be remembered, when planting a hedge, that wind considerably weakens a plant, and the cold tolerance of a bamboo is reduced considerably in windy locations, so choose species that are classed as very hardy in your area and provide initial protection if very exposed.

Hedging choices

Probably the most wind-tolerant of all bamboos is *Pseudosasa japonica*. I have seen it within a few paces of the sea on an exposed Atlantic coast, quickly increasing behind its own protection and reaching a height of about 2m (6ft). This specimen was particularly remarkable as all bamboos are very sensitive to salt spray. Just a few kilometres inland it can grow normally with only minimal leaf scorch although in very cold areas it can look unsightly.

On an exposed site a few kilometres from the sea, the Rosewarne Agricultural Experimental Station in Cornwall conducted long-term extensive tests on over 400 species and varieties of shrubs that are suitable for forming a windbreak. In their report *Pseudosasa japonica* was listed as one of the best ten species. The only adverse comment was that it tended to run at the root, which is mainly a problem on a fertile site. The easiest way to control this, which is only a once-a-year job, is to dig to a depth of one spade each side of its allotted space in winter, removing every severed rhizome as you progress. This is not an arduous task unless you once forget and the bamboo rhizomes are allowed to run into your permanent plantings.

Taller and more stately is *Semiarundinaria fastuosa*, which can grow to 9m (30ft) in a good situation, but is probably half that in an exposed site. It also can run at the root but never too much for most people, and is by far the best tall hedge plant with its lush dark green leaves and vertical form. Its only problem is that it is usually in great demand and, therefore, difficult to obtain in hedge-making quantities. Much easier to find is *Pleioblastus simonii*, which makes an excellent hedge. It looks like a smaller *S. fastuosa*, but lacks the regal bearing and rich green colouring.

All these are robust-looking species, but the very tough bamboo *Fargesia nitida* makes an excellent windbreak and has a delicate appearance unequalled in any other plant that is suitable for resisting strong winds. It is in its nature to lose all its leaves in a gale, and for the rest of the winter it looks almost dead, but if this is acceptable you will be rewarded by a flush of new, clean leaves in the spring, at a time when most bamboos are looking their worst. Other hardy mountain species such as *Fargesia murieliae* and *Thamnocalamus tesselatus* are also remarkably tough, but some do not possess the upright stature so useful for a hedge.

Many phyllostachys are suitable as windbreaks but their more open habit means that they must be planted to a greater width and are, therefore, more suitable for the larger garden. When immature, most phyllostachys are very lax in habit and are unsuitable as hedges. Unfortunately, this immature form can be permanent in cold locations. However, when they grow well, their large, vertical and well-spaced culms are ideal. *Phyllostachys bissettii* is an amazing hedge plant. It is very hardy and always looks fresh and green, even in the depths of winter. In my garden it recently came through salt-laden gales such as we had not experienced for many years. *Pleioblastus simonii* looked a mess well into summer, but *P. bissettii* hardly had a scorched leaf, and its grey-green new shoots came up better than ever in the spring.

Outside the main windbreak zone most bamboos can assist in tempering the wind with no ill effect to themselves. It is only the obviously delicate species, such as members of the genera *Drepanostachyum* and *Himalayacalamus*, that need a special sheltered spot permanently. Strong-growing bamboos are easy to transplant short distances within the garden in cooler seasons (p.51), and new introductions to an exposed garden can, therefore, be planted initially in a special location for cossetting and be moved to their permanent site when they are larger and stronger, where they will contribute to the overall filtering of the effects of the wind.

BAMBOOS AS FEATURE PLANTS

Having established what is required in the garden, and secured the boundaries, the next step in the design is to put in the main features. In addition to any structural hard-landscaping dictated by the plan, the feature planting needs to be established. That is the long-term structural planting selected for its effect and to emphasize a particular style. The plants chosen should be suitable for the area in which they are to grow. How often are large forest trees seen in small front gardens? Or shrubs that were elegant as a pot plant, now just a shapeless blob? Bamboos need serious consideration for these important spots, particularly the clumping species, of which there are many. They have the advantage that, although they grow to an imposing size, their ultimate dimensions are well known and that, after quickly reaching this height, they remain in proportion to their surroundings.

Small trees are generally purchased small because they are slow-growing: you have to wait a long time for them to achieve the desired impact, and then they keep on growing. Even a small maple may eventually become too big for its setting, and it is a hard-hearted person who could take a saw to such a specimen just because it was out of scale. Bamboos reach their mature height and there they remain, only increasing slowly in diameter. After many years they can spread out of proportion and need the size of the root ball reducing (pp.48-50). This is only necessary after a long time, however.

For an elegant atmosphere there is little better than one of the several beautiful forms of *Thamnocalamus crassinodus*, with their tiny delicate leaves, standing in isolation in a lawn or bed of low-growing plants such as English ivy. Of slightly more upright profile are *Fargesia nitida* and *F. robusta*, or *F. murieliae*, if a hardier arching plant is required. More dramatic would be *Chusquea culeou* with its large upright culms and masses of tiny leaves, or, in warmer temperate areas, one of the cultivated forms of *Bambusa multiplex* such as the beautifully variegated 'Alphonse Karr' or the smaller 'Fernleaf'.

The striking culm colours of some species can be particularly attractive in feature locations, where perhaps variegated leaves would be inappropriate. *Himalayacalamus falconeri* 'Damarapa', when growing in a sheltered mild situation, is stunning with its green, white and pink striations, but when grown in less than ideal conditions it quickly loses its colours and form. It is then best to substitute one of the many other striking plants such as *Semiarundinaria yashadake* 'Kimmei' with its yellow and green culms or the more robust outline of *Phyllostachys bambusoides* 'Castillonis' might be appropriate. In cooler areas *Phyllostachys nigra* can be used as it remains quite compact in these locations, but it can be a strong runner, as can *P. bambusoides* 'Castillonis' in warm areas.

The very hardy *Pleioblastus linearis* is liable to become invasive, but in a woodland setting with plenty of space, it is an outstanding plant and the very epitome of elegance with its arching culms and long narrow leaves. In mild situations it would be sacrilegious not to plant the stunning *Drepanostachyum falcatum* which is probably the most beautiful plant of all, with tiny leaves and culm tips touching the ground. If a more upright stature is appropriate, any of the other *Drepanostachyum* species make outstanding plants.

One of my principles is never to use a rhizome barrier to control the spread of a plant. My philosophy is that if this is necessary then the plant is wrong for the position, and in my cool area I can afford to have these high principles as only the very rampant and coarse bamboos get out of hand. There is one exception to this, however, and that is *Chimonobambusa tumidissinoda* (syn. *Qiongzhuea tumidinoda*), which is an amazing plant of elegant, weeping form with bizarre projecting nodes. In addition, it is of medium height and enjoys slight shade, so it fills a niche not exploited by many plants. Introduce this plant into the garden only after careful consideration as it has probably the most vigorous rhizome of any bamboo. I have planted it within the confines of the concrete foundations of an old shed which project slightly above ground – any rhizomes can be seen the instant they stray outside. If securely restrained inside an old oil drum or section of a large concrete drain set vertically, it makes a fine specimen plant. Large-leaved hostas or bergenias contrast well with small-leaved bamboos such as this. Also useful are the more robust-looking ferns and they, too, enjoy the shade of the location or that provided by the bamboo.

Groves and Japanese courtyards

If space permits, nothing is more evocative than a grove of giant vertical culms, or not so giant in a smaller setting. An informal path, winding between, has the qualities of a jungle track, even if the culms are no greater than 5cm (2in) in diameter. In warm zones this effect is

Phyllostachys nigra 'Punctata' is a much stronger grower than the true all-black form and can make an impressive grove.

not difficult to achieve with any of the larger phyllostachys, but in northern Europe and the cooler areas of North America most of this genus remains small, juvenile and without the vertical culms so essential for providing the atmosphere. However, *Phyllostachys dulcis* can usually be relied upon for cooler gardens. Also good are the various *Phyllostachys aureosulcata* forms and many of the hardier, recently introduced phyllostachys species. These have the added advantage that they quickly reach mature proportions. *Chusquea gigantea* (syn. *C. breviglumis*) can also reach a large size quickly, and *Semiarundinaria fastuosa* will make a good vertical grove but is slower to spread and will probably need the culms thinning to achieve the open effect.

Small areas around the house can be created in the style of a Japanese courtyard using similar plants and techniques. Never be afraid to plant a large bamboo near a house in cool temperate areas: indeed, it is essential that all gardens exploit the vertical dimension. The root systems of bamboos are not likely to cause problems like those of some trees, and the effect is stunning. A friend of mine used to have the luxury of looking out of his upstairs bedroom window through several large vertical culms of *Phyllostachys bambusoides* 'Castillonis' every morning, giving the impression of a world of green, even if the reality was usually soon to be the grey of a city. The effect was even more dramatic than from the ground floor windows. This is an effect hardly ever exploited by western gardeners, and its impact was more striking because of this. Pruning is essential in this instance. Firstly, most of the culms should be removed leaving only the imposing ones positioned to frame a view or create a foreground. Leaves and branches can be pruned to open up the view in an informal manner, or by using a stylized technique as with the pompon branches of *Sinobambusa tootsik*, so familiar in the warmer areas of Japan.

Sinobambusa tootsik is traditionally pruned to give a pompon effect which can also be produced in other bamboos. Prune branches to three internodes in the first year and to one internode in the second year.

Planting around the base of the culms usually needs to be with shade-loving species, such as *Helixine* (syn. *Soleirolia*), *Ophiopogon* or other low-growing carpet plants with moss-like characteristics. Again, English ivy can be used, or larger plants that contrast with the vertical lines and delicate leaves, such as hostas. Ferns of all sorts, including the fine-leaved species, make a good contrast with the bold culms. Taller plants should be avoided as they start to compete for attention. Gravel or pebbles are a very good alternative 'sea' through which the new culms can emerge.

One of the essentials for even a small garden is to create an element of mystery by hiding parts from view, so as to entice visitors to explore. This is easier said than done in a very small plot, where every square centimetre is put to intensive cultivation. Most plants that give a reasonable all-year-round height also spread an equal amount, taking up space that is not available. A bamboo from a medium-sized pot will be far taller than eye level, and, if planted on the inside curve of a path, will create the desired effect with practically no loss of space, particularly if backed up by other plantings or a screen. Bamboos make good screens for the washing line, compost heap or garden shed for similar reasons. They can be planted *en masse* or in conjunction with evergreen climbers supported by a vertical structure.

GENERAL PLANTING

General planting gives scope for using the many cultivated varieties that are available. One of the unique features of bamboos is the wide range of amazing culm colours that are available. Not only do they cover shades varying from white through yellow, pink, ochre, violet to black, but there are also patterns of brown and black in spots or stripes. Most of these are in a gloss finish, which makes even the normal green culm look like an emerald. In addition, there are the strange sculpted shapes of some, and combinations of shape and colour. These forms are not difficult to integrate into most gardens, creating a tropical atmosphere in which such brightness is expected. To be appreciated, however, the colours should be clearly visible, so usually it is necessary to remove poor culms and also most of the lower branches and leaves when the plants are large enough. It can be interesting to leave a few short branches and clean leaves, as the contrast between the rich dark green of the leaves and the black or yellow culms can often be arresting.

I admit that I have a great weakness for variegated plants, but I realise that this is not shared by all, and that they are not always easy to accommodate within the garden setting. Typical of these is *Phyllostachys bambusoides* 'Castillonis Inversa Variegata', an amazing plant with heavily variegated leaves in white, yellow and orange. The American form *Phyllostachys bambusoides* 'Variegata' is identical in leaf and there is nothing subtle in either of these. They are both very difficult to locate in a more natural garden setting, but who cares? There is no such problem with the more delicately variegated foliage such as *Pleioblastus chino* 'Elegantissima' or the amazingly beautiful *Sasa kurilensis* 'Shimofuri'. There are many forms between these two extremes suitable for any position or any desired effect.

In Japan low-growing bamboos are used to very good effect, both as individual plants and as carpet bedding. The *Pleioblastus pygmaeus* forms and the smaller sasas are good for this style but in a small garden can be too rampant. They have their place in a large garden or in landscaping work, but even here need to be contained by lawns, wide concrete paths, or walls. They enjoy penetrating tarmac and can easily defeat the width of a small garden path. If this feature is part of the garden design, do not forget to trim the plants down periodically to prevent them growing too tall, and this, in conjunction with a small amount of fertilizing, will improve their density and appearance. The smaller forms of *Pleioblastus pygmaeus* can almost be cut with a lawn mower. *Pseudosasa owatarii* is a real miniature gem and it even has a dwarf form, which can be used without trimming, and without qualms, but it is slow growing and, therefore, not suitable for large areas unless you are prepared to wait for results. *Pleioblastus akebono*, with its white-shaded leaves, can also be used to give a different effect, although it is slightly less tolerant and in some conditions does not perform well.

POTS AND CONTAINERS

Nothing looks more stunning than a well-grown, elegant bamboo in a pot or tub, as seen in many books and magazines. However, the reality will be disappointing if the plants are treated like most tub specimens and attention is not all it should be. Most owners are not at home during the heat of the day, go away for a few days, forget occasionally to water or feed and are happy to let a fine specimen remain in too small a pot because it is still looking good. If you are guilty of this type of neglect then forget pots and plant your specimen in the ground where it will prosper on its own.

Bamboos that are growing well are greedy plants and if they run short of nutrients they will quickly become poor in colour and form, with few leaves, and branches and twigs dying off. Only repotting and time will restore the plant to the condition where it is an asset to the garden or house. Bamboos can look like this due to just one occasion when they have run short of water. They are unforgiving plants and the roots are permanently damaged by only a short period of leaf curl. Unfortunately, unless they are growing in a generous-sized pot, their fibrous root system will soon have filled the compost completely. Coupled with their great thirst, this can lead to watering problems. Although they are hardy in the ground, all but the toughest should have their roots protected from winter cold, either by burying the pots in the garden or by bringing them into a greenhouse or a garden shed. Conversely, some of the plants from cool regions should have the pot protected from the heat of the summer sun.

So, unless you are able to give plants unfailing attention, I suggest avoiding bamboos in pots. If you like a challenge, I recommend using one of the many automatic watering systems. Also if the plants are growing strongly, as they should, divide the plant in half every spring and repot in the same sized pot after removing as much of the old compost as possible. The compost should be as open as possible and it is wise to add slow-release fertilizer to make life easier. Another common cause of failure is the compost breaking down and becoming heavy and saturated. If the plant is repotted every year this is not normally a problem.

HOUSEPLANTS

All of the *Bambusa multiplex* forms make very elegant houseplants, but in my opinion the best of them all is *Drepanostachyum falcatum* with its fine curving culms and small narrow leaves. Indoor bamboos in pots are also not the easiest of houseplants. Most require a very bright situation, out of full sunlight, and although some, such as *Bambusa vulgaris* are tolerant of dry air, they usually shed leaves and look unhappy if the pot dries out just once or the air is very dry. If your house is fully centrally-heated, it is best to keep more suitable plants, but if you have an unheated hall or other room with large windows then some bamboos can be very successful if you follow the guidelines given above. Without the humidity found outdoors plants in conservatories and houses are very prone to red spider mites, and all containers should be moved outside as often as possible during wet but frost-free spells. Watering can be a problem in hardwater areas, as most bamboos prefer neutral or acid conditions. Rainwater should be used if possible or acidifying chemicals added at regular intervals and in low concentrations.

Conservatories have special difficulties. Although light levels are no problem, ventilation is usually inadequate (permanently or through neglect) and then temperatures soar and humidity plummets, causing serious damage to bamboos.

TROPICAL EFFECTS

A tropical-looking garden is more usually seen in warm temperate regions but, with a little extra trouble, can be readily achieved in cool temperate zones, and should be seen more often. The many sensual qualities of bamboos make these plants shine above all others for this effect. In fact, I would say that bamboos are essential to produce a tropical atmosphere, and should form not only the framework, but also some of the foreground planting. I can think of no other group that combines so many jungle-like characteristics. Their bearing is reminiscent of plants of the tropics, particularly those with large culms or large leaves. The rustle of their leaves is like that of no other temperate plant, and their bright colours have an exotic lushness. What other plant combines these unique qualities with evergreen foliage and complete hardiness so that there is something to suit almost every temperate region?

Depending on your climate the list of suitable evergreen plants to complement the effect of bamboos and create year-round interest is sadly small. Assistance may be necessary from the more hardy evergreens such as laurels, aucubas and the tougher rhododendrons. The hardy palm *Trachycarpus fortunei* and its smaller form 'Wagnerianus' with distinctive upright leaves are indispensable, as is *Fatsia japonica* in both its green and variegated forms. The hardy forms of ivy, particularly those with large leaves, are also useful. You could consider the many colourful forms of *Phormium tenax* and the hardy yuccas are indispensable. Most other hardy tropical-effect plants, such as hostas, *Thalia dealbata* and the arisaemas are deciduous, and some non-hardy deciduous plants, such as *Gunnera manicata*, can be easily protected over the winter in most regions with a thick mulch. The hardy banana, *Musa basjoo*, is much tougher than most people think. I am told that in the Netherlands it can survive frozen soil conditions to a depth of 30cm (1ft) if given a covering of leaves or straw in the autumn after cutting down to ground level. It needs warmish summers to recover and prosper after these conditions but is reputed to grow up to 4m (12ft) in one season if given a good location. Its giant leaves are easily shredded in the wind so a protected site is essential.

Gardeners with warm but not frost-free gardens have a wider palette, such as the palms *Chamaerops humilis*, *Butia capitata*, *Phoenix canariensis* and others. There are also the yucca-like plants *Furcraea longaeva* and *Beschorneria yuccoides*. The otherwise herbaceous plants, such as the gingers (*Hedychium*), *Tetrapanax papyrifera*, *Musa basjoo* and other exotic species, usually keep their leaves through the winter in these zones.

In the summer both of these zones can benefit from an injection of tropical plants that have been overwintered in a greenhouse kept just frost-free. These may have been either kept in pots all summer or planted out and lifted and repotted in the autumn. Apart from the very exotic forms of cannas there are various cacti, succulents, bromeliads, the grasses (*Cyperus papyrus* and the variegated form of *Arundo donax*) and many other species that prosper in our cool summers if given some winter protection. Too many

Sasaella masamuneana 'Albostriata' (left) and *Pleioblastus viridistriatus* (right) in a pot with other plants.

greenhouses are left empty in the winter when they should be crammed with such plants, making the most of that bit of heat that keeps the temperature above freezing and tides them over their winter dormancy.

Municipal gardens could make more use of this type of garden. They have the facilities and staff to undertake large bedding schemes of annuals, which are often raised under much warmer winter conditions than the perennials, and how much more appropriate the end result would be, particularly for holiday areas. Similarly the many hotels in cool locations invariably neglect to exploit the tropical effect that can be achieved through bamboos and similar planting. The approach to a hotel is free advertising space. It sets the tone for not only the hotel but also the whole holiday. Why then do they mainly rely on mundane planting schemes and cheap labour?

LARGE GARDENS

Those lucky enough to have large gardens with small areas of water should consider locating bamboos where their reflection can be appreciated. The effect achieved is much in the style of a weeping willow, but a bamboo is usually more in scale and can be selected not to outgrow its welcome. All bamboos enjoy being positioned a few feet above water level where they can have constant ground moisture and air humidity without their roots being in a bog. Arching species, and in particular clumping varieties, are usually most appropriate, in particular *Fargesia murieliae*, *Fargesia utilis* or the various elegant *Thamnocalamus* or *Himalayacalamus* species and forms. One of these planted to one side of a bridge and reflected in water would have all the elegance of a Monet painting. Lush leafy water plants, such as *Rheum palmatum*, *Thalia dealbata* and water lilies, associate and contrast effectively with the upright delicate bamboo form. Most bamboos, even the more robust species, look good on the banks of a very large pool.

Large gardens can also utilize bamboos effectively on the edge of informal woodland where they will add a touch of grace and tranquility, and appear much as they would in the wild. Specimen plants can be used at the ends of vistas or either side of a path to frame a view. Large upright clumping bamboos can also be used at the back of borders or in the centre of island beds to add height. Many of the spreading species are useful for stabilizing river banks, cliffs and exposed earth, provided their invasive potential is understood (p.54).

SMALL GARDENS

Smaller gardens and town gardens can also benefit from bamboos used to add height, as they take up little space, cast little shadow if upright species are selected and do not rob the surrounding plants of nutrients or water. They associate well with almost every other plant but look particularly effective with those that have lush foliage or that complement or contrast with their delicate foliage or robust form. The list is endless – ligularias, kniphofias, ferns, hostas and grasses are but a few. Very small gardens should use bamboos, where height is needed, in preference to trees, which bring their own problems that can be particularly acute in very small plots. Bamboos have an unfortunate reputation of spreading like weeds, but there are dozens of species that remain in a compact clump for many years if trouble is taken to select the right plant, rather than those that are suitable for mass production by the thousand to satisfy the garden centre trade. Any of the hardy *Fargesia*, *Shibataea* or *Chusquea* species or varieties would be useful in a small garden setting. The larger species are good as background or specimen plants and the smaller ones for general planting. In warmer locations, small *Bambusa* and all *Drepanostachyum*, *Himalayacalamus* and *Thamnocalamus* can also be used.

One type of garden which is surprisingly successful is one, or an area within a larger one, that is predominantly given over to bamboos, as can be seen at the Royal Botanical Gardens in Kew. I have an area devoted to bamboo species and forms. Originally this was for bamboos growing within other plants around an informal lawn but, as my interest grew, bamboos began to predominate. I had no restraints as the area is away from the house and, in time, most of the other shrubs disappeared along with the lawn. Now informal paths wind through large culms, within a carpet of leaf litter or dwarf bamboos. It matters not that every few paces the species or form changes, for they are all essentially very similar, and these variations add variety and interest without introducing discord. I have added a seat or two (an essential component in any garden) at special spots that have been produced spontaneously. To sit within a bamboo grove is now my retreat from the pressures of life, and friends – both gardeners and the unconverted – delight in joining me.

Drepanostachyum falcatum enjoys a woodland setting and can grow quickly. This specimen is only a few years old.

6

AN A-Z OF BAMBOOS

PLANT CLASSIFICATION

Linnaeus (1707–1778) methodically classified the whole living world into the dual name system that is used today. The first (or generic) name can be likened to our surname, except that it is unique to one family (or genus). The second (or specific) name is given to one plant type (or species) within that genus, and can be likened to our first (or Christian) name. The specific name has a descriptive function and can also be used for species that belong in other genera. Although it is tempting to use common or ethnic names these are not recognized universally and, in any case, many species names are very descriptive and so should be easy to remember. In informal conversation it is common to use the specific name only, and this helps to avoid confusion as it is almost always the generic name only that is subject to changes. What could be easier, more descriptive and more readily understood in this international world of bamboos than *Pseudosasa japonica* (the *Sasa*-like plant from Japan) for instance? The alternative is the arrow bamboo, a meaningless title outside its country of origin. But if the latin is still a problem what about remembering just 'japonica'?

The divisions of classification of a lower order than a species are divided into naturally occurring botanical variations and cultivated varieties. Subspecies (subsp.), variety (var.) and forma (f.) are botanical divisions and are always given Latin names. Subspecies and varieties have definite but minor morphological characters that distinguish them from the standard type of the species. A subspecies usually is found in a definite geographical area, such as the western end of the growth range of the species. A variety usually has a more broken distribution, for example in drier habitats, where it might be hairier, in order to resist browsing pressures. A forma is a still narrower group of plants, for example very erect forms of a particular conifer. Examples of these botanical types are *Thamnocalamus spathiflorus* subsp. *nepalensis*, *Chusquea culeou* var. *tenuis*, *Chimonobambusa macrophylla* f. *intermedia*. Theoretically, it is possible to have a name such as *Bambusa vulgaris* subsp. *occidentalis* var. *hirsuta* f. *erecta*, but in reality no such plant exists.

Cultivated plants (cultivar or cv.) have a slightly different system of names, with the cultivar name used either without a species, like *Bergenia* 'Snowstorm', or more frequently added to a species name where the appropriate parent is known: *Agapanthus orientalis* 'Wavy Wavy'. They are not in Latin, and are given single quotes and an initial capital letter. A cultivar is genetically the same as a botanical forma, but is published as a cultivar and under different rules. In bamboos, however, the origins of most garden forms are unknown. For instance, the form 'Castillonis' of *Phyllostachys bambusoides* is recorded as being introduced into Japan from Korea towards the end of the sixteenth or the beginning of the seventeenth century. The wild green *Phyllostachys bambusoides* originated in China so who can say for certain where or how this variation occurred? Most bamboos that differ from the type have originated in the wild or by pure chance in cultivation and are almost never artificially induced. Often, large colonies of these forms, such as *Bambusa vulgaris* 'Vittata', are to be found growing wild and are not

Phyllostachys bambusiodes is a strong-growing species but this specimen has had all its distinctive robust lower branches removed to enhance the appearance of its culms.

dependent on mankind for their preservation. Therefore with bamboos these academic divisions are often not clear and, just as with the naming of bamboos, differing opinions exist. In this book the word 'forms' is used to cover any variation from the type plant whether originating naturally or in cultivation.

NAMING POLICIES

In recent years bamboos have been quite extensively reclassified as our understanding of them has improved. Temperate bamboo species with circular hollow culms have been subject to most of the constant and confusing name changes. This was already the case when A. H. Lawson wrote *Bamboos: A Gardener's Guide to their Cultivation in Temperate Climates* in the late 1960s but he wisely avoided confusion at that time by grouping them under the broad title of *Arundinaria*. Quite distinct horticultural as well as botanic differences can be seen between the species that formed his subdivision, so there is good reason now for not adopting Lawson's logic, but instead recognizing several smaller genera. The names used in this book largely comply with the Royal Horticultural Society's database at the time of publication. This database has been compiled by a team of eminent scientists drawn from Kew, the RHS and several universities. They have a policy of retaining the old name until a clear consensus is reached upon any changes. This is important as it helps to maintain the stability desirable in horticulture. For this reason, however, there may be other opinions upon classification in operation. One exception to the following of the database is *Chusquea gigantea* (was *Chusquea breviglumis* or *Chusquea aff culeou*) which has been published very recently, and a couple of others are named on the advice of Kew.

IDENTIFICATION FEATURES

Many features of bamboos are variable or transient, such as leaf form, leaf colour, plant form and plant height. Where these are given in the plant descriptions they are not meant to be used for identification purposes but as a guide for plant selection, or to reinforce an identification based upon permanent features. The process of plant identification should begin by determining the genus from culm and branch features at eye level (1.5–1.8m/5–6ft) in conjunction with the mature rhizome type. Sometimes the species name can then be established within the genus by a distinguishing feature, but often it is necessary to examine the new shoots before it can be identified with any certainty.

Culm dimensions given in the plant descriptions are the normal maximum when the bamboo is grown in ideal conditions. Height is, therefore, usually less in species with arching culms or when the plant is growing in average garden conditions. Leaves are usually very variable, not only with growing conditions but also on one plant. Terminal leaves are often very large and leaves on older culms are often much smaller than the average. The leaf dimensions stated try to be an average of all these variables and to represent that found most frequently on a particular species.

Temperatures stated are also only an approximate guide as the hardiness of any plant is dependent upon so many other factors, including its age, wind conditions and the quality of its growth during the summer. I have assumed that some winter protection, such as heavy root mulches or temporary screens, will be used in very cold locations. One of the major attributes of bamboos is their unique evergreen winter qualities, but it should be understood that most bamboos will have lost this important feature before temperatures reach the stated minimum. This is particularly so in a windy area. Some species are recent introductions of unproven hardiness. For these, minimum temperatures have been estimated based upon other available information. The hardiness figure is followed by a question mark to indicate that it is an estimate.

AMPELOCALAMUS

A small genus adapted to scramble up branches and undergrowth. Plants are found below 1500m (5000ft) in south China and the Himalayas. Culms are often rough and striated with prominent corky rings at the nodes to help give purchase on branches of trees. Culm sheaths often have finely fringed margins, and the multiple branches have strong knee-like joints, again to assist with climbing. They are very attractive pendulous bamboos but are not hardy outside in most regions and cover a large area if grown in a greenhouse.

Ampelocalamus scandens

Min. 0°C (32°F)?, Culms 10m x 8mm (30ft x ⅜in), Leaves 5 x 1cm (2 x ½in). A hanging species with very prominent spider-like oral setae. It is known to be growing in a few protected places in southern Ireland and southern Europe. It has very extended, small-diameter culms with larger leaves on the terminal growths. In cold locations its growth is usually shorter and stiffer. Seed was introduced from Yunnan into the USA a few years ago and distributed to a number of locations worldwide.

ARUNDINARIA

This once-large genus has now been reduced to just a few species. To the horticulturist the genus is identical to *Pleioblastus* and a detailed description is given under that heading.

Arundinaria fangiana (*Bashania fangiana, Gelidocalamus fangianus*)

Min. -20°C (-4°F), Culms 1m x 7mm (3ft x ¼in), Leaves 5 x 1cm (2 x ½in). This diminutive plant is one of very few small pachymorph species available to the gardener. It was introduced under the name 'Tung Chuan 2' and started to flower shortly afterwards, therefore little is known about its cultivation or its mature characteristics. The very small seeds were difficult to find but they germinated easily, often on the surface of the soil surrounding the flowering plant. Seedlings were also very small and slow-growing so it will be some years before it is again available. Culm sheaths are fringed with hairs (ciliate) and there are no auricles or oral setae. Branches are multiple and have 2–4 light green leaves per twig.

Arundinaria gigantea

Min. -20°C (-4°F), Culms 9m x 4cm (30ft x 1½in), Leaves 20 x 2.5cm (8 x 1in) (very variable). This is the only bamboo species to be found on the North American continent, where it once formed vast thickets or 'canebreaks' around the water courses of the southern states establishing a unique ecosystem. Until relatively recently it was an important grazing plant but it is now quite rare in the wild due to extensive removal by fire or over-grazing. It is reported as being very slow or impossible to re-establish after clearing because of invasion of other species.

As a garden species it is very variable but, even so, it is easily identified by several very distinct features described below. To most gardeners it is not special enough to warrant a prominent position and not tough enough to be used as a hedge, and, consequently, it is rarely cultivated. A well-grown plant is impressive and unusual, however, and could be positioned in a damp open glade in a warmer location.

Its culms are thin-walled and circular in cross section. They are light green quickly changing to dull yellow. Initially, it has one central branch with several smaller branches arising from between its compact basal nodes. The number of branches increase with age to form multiple, short, upright bunches. New shoots are deep purple and very soft. Culm sheaths have a purple hue when young and are persistent. When growing well it has a superficial resemblance to a large *Pseudosasa japonica* and when in poor conditions to *Pleioblastus humilis*. It is easily and quickly distinguished from both of these species by its consistent pale yellow-green leaves (only found when cultivated in the cooler regions), which vary considerably in size on one plant and are very soft, and its soft, fragile new shoots. Conversely, at the same time, the leaves look coarse.

A naturally occurring form or subspecies called **'Tecta'** seems to be more adapted to damp soil conditions with peripheral air channels in the rhizomes, and also varies from the species by having wine-coloured flowers (those of the species are straw-coloured). Some also claim that it is smaller with longer branches, but none of these feature give it any horticultural distinction. In the cooler areas, the growth of both plants is greatly curtailed and then any distinction based upon height is academic.

BAMBUSA

There are several *Bambusa* species that will withstand a few degrees of frost but the genus is essentially tropical. A few of the smaller species can be cultivated in a large pot and brought indoors during the winter. Others are suitable for cultivation in warm locations outdoors, but their performance is dependent as much upon hot summers with high light intensity as upon warm winters. If you try them ouside, it is often very late in the year before new growth is made. Also, the first flush of new leaves is sometimes white or white on green, which is a sign that the plant has been stressed through the winter. Unless you like a challenge it is usually advisable to choose species from more temperate genera.

The genus is identified by its distictive culm sheath blades, which are very broad in relation to their length.

PLATE VII

Bamboo leaf colour

Pleioblastus chino
'Kimmei'

× *Hibanobambusa tranquillans*
'Shiroshima'

Semiarundinaria yamadorii
'Brimscombe'

Pleioblastus chino
'Elegantissima'

Scale approximately half lifesize

They are also broad at the top and the blade plus sheath is very prominent. The branches are similar to those of the genus *Himalayacalamus*, but with one prominent central branch and several smaller branches increasing in number up to about 30.

The rootstock of the genus is pachymorph but on some species is quite open in form. The culms are circular in section. The leaves are not tesselated and new shoots generally form late in the year.

Bambusa multiplex (*Bambusa glaucescens*)
Min. -7°C (20°F), Culms 5m x 3cm (16ft x 7½in), Leaves 10 x 3cm (4 x 1¼in). This elegant small species is one of the most variable cultivated plants and it has numerous desirable forms. Its close-packed culms can be kept much smaller than the size given above by repeated division. For this reason it makes a very good pot plant; several of the cultivated forms are ideal for pots.

It has 5–10 leaves per twig. The leaves have distinct, abruptly tapered ends making a paddle shape. The culm sheaths have small auricles with pointed ends that curve outwards and the culm sheath blades have prominent lines of dark hairs. The plant emits a pleasant perfume, which is instantly noticeable in a greenhouse or room, but it is not unique among bamboos in this respect. The species is not often cultivated in temperate zones but the following forms are popular.

'Alphonse-Karr' has attractive bright yellow culms marked with random dark green stripes. This coloration, in conjunction with its elegant form and thin-walled circular culms, makes it instantly recognizable. The similar *Himalayacalamus falconeri* 'Damarapa' can be distinguished by the white variegations on its culms.

'Riviereorum' is a delightful dwarf form, ideal for pot culture and reputed to be the hardiest of the forms. It grows to about 2m (6ft) high and its leaves are only about 3–4cm (1¼–1½in) long and arranged in neat rows of 12–23 per twig. A small group of this form has been growing outside at Kew gardens for some years but has not increased greatly in size. It can easily be confused with the form 'Fernleaf' but is a smaller, more desirable plant with solid culms.

'Silver Stripe' can be seen in the West, and it is sometimes sold under the label 'Argenteo-Striata'. Not an outstanding form, this is like the type plant but has white stripes on most of the leaves and a few fine lines on the culm internodes.

Bambusa ventricosa (Buddha's belly)
Min. -7°C (20°F), Culms 2.5m x 1cm (8ft x ½in), Leaves 12 x 1.2cm (5 x ½in). This bamboo is generally purchased as an imported bonsai with distorted and compressed internodes and only a single, distorted branch at each node. It will consistently fail to reproduce these features for its new owner. In its native China selected plants can frequently be seen as an ornamental feature and freak culms seem to be relatively easy to produce there, but I have never seen or heard of a single distorted culms being produced in the colder regions of the West.

Bambusa vulgaris
Min. 0°C (32°F), Culms – can be very large, Leaves 12 x 1.2cm (5 x ½in). Probably the most common bamboo in the world because of its easy propagation from culm-node cuttings. It is tolerant of dry conditions, which makes it suitable for pot culture, but its usual response is to replace all its leaves if the pot accidentally dries out, which is not popular if it is used as a house plant. This species also emits a delicate perfume.

'Vittata' is the form that is usually grown. An outstanding bamboo, it has substantial brilliant yellow and green culms making it instantly recognizable. It is readily distinguished from the similarly coloured *B. multiplex* 'Alphonse-Karr' by its more robust appearance and the wider spacing between culms. It is not so effective as a pot plant for cold locations as the colours of the new culms are easily stained brown by temperatures near freezing. Sometimes culm cuttings of this form can be seen for sale to the house-plant market. Freshly cut culms about 5cm (2in) diameter are imported and cut into sections with at least two nodes. The sections are potted vertically in greenhouse conditions where the lower nodes produce roots and the upper node forms an umbrella of trimmed branches. If you want one of these to live more than a few years, select a specimen which has not only rooted but also shows signs of rhizome activity. This can be seen by new culms forming away from the old culm. Some culm cuttings grow roots but no rhizomes and are only able to produce branches from each node.

'Wamin' produces compressed internodes, but although this feature is widely seen in specimens growing in hotter climates, it is very rarely produced in cultivation in temperate areas.

BASHANIA

This small genus is very closely related to *Arundinaria* and, like that genus, is indistinguishable horticulturally from the genus *Pleioblastus*.

Bashania fargesii (*Arundinaria fargesii*)

Min. -25°C (-13°F), Culms 10m x 5cm (30ft x 2in), Leaves 15 x 2.5cm (6 x 1in). This very tough species has leptomorph rhizomes and should be restricted to utilitarian purposes. I have it as a windbreak in a very dry site where little else would grow. It never shows signs of distress and has even colonized the area. When growing well, it is tall with widely spaced culms and would form a good grove in cool growing areas. It is very vigorous, however, and needs control in most climates.

The culms are circular in cross-section, thick-walled and grey-green. The brown-haired sheaths are soon shed. At first, each node has three branches, later multiple branches develop. The leaves are hard and tough.

When small, this species can be confused with *Pseudosasa japonica*, which is distinguished by single upright branches and persistent culm sheaths. *B. fargesii* also grows much taller and larger, and is also much more active at the roots with wider-spaced culms. In very cold regions it is a good substitute for *Pseudosasa japonica*.

Bashania qingchengshanensis (*Arundinaria qingchengshanensis*)

Min. -20°C (-4°F)?, Culms 4m x 1cm (13ft x ½in), Leaves 25 x 3cm (10 x 1¼in). This species is a very new introduction and as yet there has been little experience of mature plants. It should combine the assets and disadvantages of *B. fargesii* with a smaller stature, and a less stiff style of growth. The culms are solid or nearly solid and marked grey-yellow below the nodes. It has persistent sheaths, the bases of which can remain attached to the node long after the sheath has gone. They are thick and hairy with no auricles. There are 5–12 branches per node bearing hard, long lanceolate leaves that tend to wither at the ends during winter. This is not a particularly good-looking species but could have a position in the garden in difficult places or as a hedge.

The species is found at lower elevations than *B. fargesii*, which suggests the given minimum temperature. However, Max Riedelsheimer of Stockdorf in Germany, reports that it is at least as hardy as its larger relative and came in to growth early, with less leaf damage.

CHIMONOBAMBUSA

Members of this Chinese genus generally enjoy damp conditions, slight shade and good humidity levels, and have leptomorph rhizomes that can be invasive in good conditions. This is the only genus in cultivation that has leptomorph rhizomes and (when established) normally grows new culms in the late summer or autumn. Additional identifying features are that it has prominent nodes and there is more than one branch per node. Branches develop in the spring and dormant air roots are often found on the lower nodes. The culm sheaths are thin and papery with very small sheath blades. The culms are circular in section or have a tendency to be four-sided. Divisions are often slow to establish as there are usually only a few roots on the base of each culm. Newly established plants can produce culms during any cool, damp season.

Chimonobambusa macrophylla f. intermedia

Min. -20°C (-4°F), Culms 2m x 1cm (6ft x ½in), Leaves 5 x 1cm (2 x ½in). This small species was introduced recently and is still rare in cultivation, when it usually grows to only about half the height given above. It requires a high relative humidity. The nodes are very prominent, the auricles and oral setae are absent, and there are multiple branches with 2–4 leaves per twig. It has flowered in a few locations and the seeds are very large and dark green. They germinate readily – in one case in the polythene bag in which they were stored.

Chimonobambusa marmorea

Min. -20°C (-4°F), Culms 3m x 2cm (9ft x ¾in), Leaves 10 x 1cm (4 x ½in). A native of Japan, this species was first introduced into French gardens by M. Latour-Marliac in 1889 and into the rest of Europe shortly afterwards. It is very invasive and difficult to control when growing strongly. When it grows tall in conditions it enjoys, it is an impressive plant with the weight of lush foliage causing it to arch gracefully. It is normally a second-rate garden plant, however.

The leaves are hairless; the upper surface is shiny and medium green. New shoots are pale green, mottled brown and white, and tipped pink. The fresh sheaths are similarly mottled, thin and persistent. The sheath blade is very small – only 1–3mm long. There are usually three branches per node with one longer than the

PLATE VIII

Bamboo leaf colour

Scale approximately half lifesize

Phyllostachys edulis 'Bicolor'

Phyllostachys bambusoides 'Castillonis Inversa Variegata'

Phyllostachys bambusoides 'Variegata'

Pleioblastus viridistriatus 'Chrysophyllus'

Sasaella masamuneana 'Albostriata'

others. Early branches form in a distinctive herringbone pattern. The culms are thick-walled with prominent nodes and often have air roots on the lower nodes. It is easily identified from others in the genus by its marbled new culm sheaths, its upright, bunched leaves and branches, in combination with its small size.

This species has been reported as flowering on a number of occasions but over the last sixty years continuous partial flowerings have been recorded and perhaps this is its normal style. The flowers are a distinctive dark colour and the seeds are large and prominent, and also dark.

Chimonobambusa marmorea is easily recognized by its small size and upright, bunched branches and leaves.

'Variegata' is like the species and can be a fine plant when well grown but it is usually unimpressive. The culms are yellow with a few green stripes, although this is lost on small plants. There are a few white stripes on the leaves, but not enough to be noticeable. The yellow of the culms turns red in strong light and this can be outstanding if the plant is a reasonable size. This form has also flowered and the resulting seedlings are either all green or all white.

Chimonobambusa quadrangularis

Min. -11°C (12°F), Culms 7m x 4cm (22ft x 1½in), Leaves 14 x 1.5cm (5½ x ½in). Thriving in deep woodland shade, this species has very distinctive tall, upright growth. The culms are thick-walled, tough and an unusual matt grey-green. They have longitudinal ridges and in larger culms are obviously four-sided. The nodes are pronounced. The upper nodes have 3–5 branches. There are no branches on the lower nodes but there are air roots, which act as thorns (gloves should be worn when pruning). The sheaths are straw-coloured with purple blotches and fall early.

This species is not very hardy in drier regions, but in moist, cool locations it can become large and very invasive. It originates from China but is extensively grown in Japan and most other neighbouring counties that have a suitable climate. It was introduced into Europe during the last century where its cultivation is limited to the milder, wetter regions. In spite of all this cultivation, it has never been recorded in flower until one flowering culm was found in Heligan Gardens in Cornwall a few years ago. This was cut as a botanic specimen but no further flowers have been found.

This plant is not easy to propagate when large, as there are few roots on the base of the widely spaced culms. It is best to take rhizome cuttings in the spring.

'Suow' is a fine cultivated form that has soft yellow culms with a few green stripes, and a few variegations on the leaves. The colouring is best on large culms grown in a shady situation. **'Nagamineus'** is similar but has a broad green stripe above each branch.

Chimonobambusa tumidissinoda

(*Qiongzhuea tumidinoda*)

Min. -11°C (12°F), Culms 6m x 2.5cm (20ft x 1in), Leaves 10 x 1cm (4 x ½in). This very elegant and fascinating species has recently been transfered to the genus *Chimonobambusa* but differs from most *Chimonobambusa* in that its new culms arise in the spring and its culm sheath blades are quite large.

Although only recently introduced into the West by Peter Addington, there are several plants in cultivation well over 3m (10ft) high and still increasing. It was thought to be not very hardy, but experience suggests that it is at least as hardy as *C. quadrangularis*. Its outstanding feature is its extremely pronounced culm nodes that result in it being in great demand in parts of China for walking sticks. Equally impressive are its gentle curving culms and small long leaves making it probably one of the most elegant cultivated species. It usually has three branches per node and 3–4 leaves per twig. The leaves are light green in the sun but darker when grown in the shade.

It probably has the most vigorous leptomorph rhizome system of all cultivated bamboos and can be seen with numerous runners radiating several metres away from the parent plant. I have even seen the rhizome cross a small stream producing water roots in the process, a feature not experienced with any other bamboo. These rhizomes are particularly dangerous in a small garden for they do not immediately send up culms and may travel many metres before they are detected.

CHUSQUEA

This is a very large and distinctive genus from South America. Its solid culms with a central core of pith are an instant aid to identification. The feature is not unique to the genus, as parts or sometimes whole culms of other species can exhibit this structure, but in conjunction with a study of the branches, the genus is easily recognized. Other than *Chusquea*, all multi-branched cultivated bamboos branch and re-branch from close-packed buds. This branching usually takes place at the greatly compressed basal nodes of the first or primary branch, so that the branch complement arises in close proximity. *Chusquea* has multiple branch buds, often radiating well around the circumference of the culm, which gives rise to their distinctive appearance. There is usually one large bud, which produces the main branch, surrounded by many small buds producing short branches. The main branch often emerges some time after the many smaller branches. All *Chusquea* species presently in cultivation have pachymorph rhizomes, and auricles and oral setae are always absent.

In the wild, they often take advantage of forest clearance just as they do when trees fall naturally. After logging operations, large dense areas of bamboo can be found, particularly on sloping ground. They are plants of the high Andes with some found almost up to the permanent snowline. There are probably many interesting species, subspecies and forms awaiting introduction from the large botanically unexplored temperate areas in these mountains.

The species in cultivation grow naturally in swirling mists in light woodland or sometimes in the open and they resent high temperatures and dry air. Most temperate species rarely experience temperatures above 25°C (77°F) and probably 20°C (68°F) is the top of the comfort zone. This intolerance of hot conditions by all cultivated species, and cold conditions by some, causes the majority of the cultivation problems. A damp, cool maritime climate suits them well, and some are very hardy and very wind resistant when given these conditions. The larger species are remarkably quick growing when well established.

There are some differences of opinion between scientists about the classification of some members of the genus into species, subspecies or forms. Horticulturally they are distinct and are, therefore, listed mainly as species. *American Bamboos* by Judziewicz, Clark, Londono and Stern has been used for the latest nomenclature, with the exception of *C. gigantea*. Most of the information given here regarding growing conditions found in the wild is based upon the observations and records of Friedrich Schlegel-Sachs given in the newletters of the European Bamboo Society (Great Britain) November 1997 and August 1998.

Chusquea andina

Min. -15°C (5°F), Culms 1m x 1cm (3ft x ½in), Leaves 10cm x 8mm (4 x ⅜in). A dwarf species with stiff upright growth, short upright branches and narrow leaves. It grows above the timber line in the Cautin to Chiloe Provinces and probably has similar cultural requirements to *C. culeou* combined with a greater tolerance to cold. *C. argentina* is very similar but comes from the General San Martin region in Argentina, which has a dry, cool climate. It probably combines the best temperature- and drought-resistant properties of the genus.

Chusquea coronalis

Min. -4°C (25°F), Culms 7m x 3cm (22ft x 1¼in), Leaves 3cm x 2mm (1¼ x 0.1in). A beautiful tropical species suitable for pot culture in temperate regions. It originates from slightly shaded locations in Central America but, in cooler areas, requires high light intensity and summer heat to prosper. It looks better with the main branches removed.

Chusquea culeou

Min. -18°C (0°F), Culms 6m x 2.5cm (20ft x 1in), Leaves 10 x 1cm (4 x ½in). A very variable species with almost every specimen having different characteristics. This striking plant was introduced into England from Chile in 1890; seed followed in 1925 and gave rise to many much sought-after clones. Much more recently further batches of seeds were collected. These gave rise to clones with other distinct characteristics and these form the majority of the plants being sold today. At the time of writing the older clones are in full flower and large quantities of seed have been collected. Raising the seedlings is not proving easy, however.

In spite of its variability, this species still cannot be confused with any other bamboo. Typically, it is tall with thick, solid culms, which are sometimes rigidly upright, but usually arch gently under the weight of the foliage. The multiple branches extend two-thirds of the way around the culm. They are fine, and normally the central large branch is lacking. The hundreds of small leaves form distinctive 'fox brushes'. The rhizomes are pachymorph, but when the species is grown in a woodland location, it usually produces larger, but fewer culms with greater spacing in between. The only other bamboo with which it can be confused is *Thamnocalamus tesselatus*, but only then when it is not growing well. A quick examination of the features of the genus will resolve the identity.

In maritime situations it is very hardy and very wind resistant but it is not so tolerant in less favourable locations. Divisions can be taken in the autumn if your autumns are long, cool and damp, or very early spring in other areas, but success is very difficult to achieve in anything less than ideal conditions. Those who have good results provide cool, humid conditions in a well-ventilated greenhouse or tunnel. Large plants are even more difficult to propagate and established plants should never be moved. It is recorded as originating

Chusquea gigantea (p.96) reliably makes a grove of widely-spaced culms in Britain and Europe.

PLATE IX

Bamboo branch formations

Chusquea gigantea has multiple branches and one large one

Sasa has single branches

Semiarundinaria normally has three branches

Himalayacalamus has multiple branches

Scale approximately half lifesize

Drepanostachyum has multiple branches (more than *Himalayacalamus*)

× *Hibanobambusa* has two or three branches

Chusquea valdiviensis has multiple branches and one very large one

Pleioblastus simonii has branches that make a triangular shape at their base

The rare *Chusquea cumingii* is very distinctive with its short, stiff branches and very small leaves.

from areas with loamy acid soil and high rainfall most of the year, but as it grows over a wide area this may not be important.

There are many distinctive forms of C. *culeou* from the old seed introductions and these usually bear the name of their location: 'Saville Gardens', 'Ness' and so on. **'Tenuis'** is the only one that is sufficiently different to be listed here. It is a very distinctive dwarf clone not known in the wild, but there are vast areas of the Andes waiting to be explored botanically and, therefore, it is reasonable to question its identity as it bears similarities in growth to other scrubby high-altitude *Chusquea* species. 'Tenuis' grows to about 2m (6ft) high and has larger leaves than the type. Branches and leaves are upright and the plant spreads slowly at the root so that old plants can cover a large area. (It is not suprising that earlier enthusiasts considered it a separate species. Unfortunately they called it 'breviglumis', which is the name of a different species not in cultivation, and the name also now used, inaccurately again, for C. *gigantea*.) It has recently flowered and most plants are now dead. Good seed has been obtained and it is hoped that some new plants will be available despite it proving to be difficult to raise in most locations. Seedlings are distinctly different to those of C. *culeou*.

Chusquea cumingii

Min. -10°C (14°F)?, Culms 1.5m x 1cm (5ft x ½in), Leaves 5cm x 6mm (2 x ¼in). This small species grows in the Aconcagua to Arauco Provinces up to 1300m (4300ft) in a Mediterranean to warm temperate climate. It enjoys full sun with low rainfall and grows in alkaline loamy clay. Although the species was recorded as being introduced into England in the early part of the twentieth century, no plants from this period remain, so their identity is still in doubt. Current plants in cultivation have been introduced recently and are proving to be reasonably hardy. The plant has very short, slightly clambering culms with short, bunched, distinctly stiff branches and tiny, narrow, glaucous leaves. This is a very desirable dwarf garden plant for those who can provide an oceanic climate with temperatures not too cold in the winter or too hot in the summer.

Chusquea gigantea (*Chusquea aff culeou, Chusquea breviglumis*)

Min. -15°C (5°F), Culms 15m x 5cm (50ft x 2in), Leaves 12 x 1cm (5 x ½in). This is a plant with a serious identity crisis. The plant described here was originally considered to be a regional form of C. *culeou*. It is currently grown under the incorrect name of C. *breviglumis*, but the true C. *breviglumis* is a smaller plant not known to be in cultivation. French botanist Jean-Pierre Demoly has just published the name C. *gigantea* for it and, although probably not universally accepted, this is ideal for the horticulturist. To add to the confusion, before about 1985 any printed references to C. *breviglumis* were normally relating to a dwarf form of C. *culeou*, a very different plant now called C. *culeou* 'Tenuis'. Many people were unaware of this name change so any reference to C. *breviglumis*, even to this day, needs checking. Fortunately, it is easy to establish which plant is being discussed as there is little similarity between them.

Notwithstanding identity problems, this is an outstanding plant, in great demand. It is very upright and tall and has several distinct features. Compared to C. *culeou*, the pachymorph rhizome system is open, with probably 30cm (1ft) between culms, and, unlike most other bamboos, which have rhizomes near the soil surface, the rhizomes of this species grow very deep.

When given the damp oceanic climate it enjoys, this is a very wind-tolerant plant. Reports from growers with drier continental climates indicate that it is not as hardy as C. *culeou* in these areas. It is also not so sensitive to propagation by division. The problem with dividing a mature plant of this species is the large extended rhizomes, which will be found attached to even a single culm division, and its depth in the soil – a very large pot should always be available before starting this work.

Chusquea montana

Min. -15°C (5°F), Culms 1.5m x 1cm (5ft x ½in), Leaves 7 x 1cm (3 x ½in). This small species is distinguished by its swollen nodes and short branches. Seedlings of a small form of C. *culeou* have been introduced into Europe under this name so there is some confusion with the identification.

The species is found in the Valdivia Province to the Queulat National Park, in cool temperate conditions with high rainfall. It is found in *Nothofagus* (southern beech) forests near the timber line at 1300m (4300ft), and grows in loamy, acid volcanic soil. It enjoys some shade.

C. *montana* var. *nigricans* is smaller with glaucous leaves. It grows near bogs in cool temperate parts of Valdivia to Chiloe Provinces and is probably even more hardy, growing above the timber line in full sun. The climate has very heavy rainfall and the soil is a very acid peaty clay.

Chusquea quila

Min. -10°C (14°F)?, Culms 4m x 1.5cm (13ft x ½in), Leaves 5cm x 6mm (2 x ¼in). An attractive bamboo but not easy to place in anything but a wild-type garden, this is a climbing or, more accurately, scrambling plant with thin culms that emerge from the compact rootstock almost parallel to the soil and obtain support from the surrounding vegetation. It has short, stiff, bunched branches and very small dark green leaves.

It originates from Valparaiso to North Cautin Provinces in an area of Mediterranean climate. Rainfall is low and it is found in full sun to half shade in acid loamy clay conditions. This species was recorded as being introduced into England in the early years of the twentieth century but as none of these plants remain their identity cannot be confirmed. Seeds of an open and upright form of C. *culeou* have been introduced into Europe under this name but these are easily distinguished from the true species.

Chusquea uliginosa

Min. -10°C (14°F), Culms 5m x 2cm (16ft x ¾in), Leaves 8 x 1cm (3 x ½in). A clambering species that likes warm wet conditions, C. *uliginosa* originates from Valdivia to Chiloe Provinces in temperate areas and shallow, acid loamy volcanic soil on shingle of glacial origin. It has arching culms and enjoys full sun in cultivation.

Chusquea valdiviensis

Min. -7°C (20°F), Culms 15m x 2.5cm (50ft x 1in), Leaves 12 x 1cm (5 x ½in). This is a climbing species with amazing vigour from the Arauco Province to Palena, growing in temperate conditions in full sun to 40 percent shade. It grows best in evergreen forests where the soil is an acid volcanic loam. When it was introduced into Europe by seed it came with a dire warning from the collector. He stressed that in its native country, it swamped all vegetation 'like a giant bramble', rooting from the tips and forming large impenetrable areas. It is easy to recognize with its substantial arching culms that rise into the tree canopy and descend to the ground in a never-ending, very fine, whip-like terminal growth. Initially the multiple branches are short, rigid and backward-pointing. These act as grappling irons upon anything that comes in its way. When they are secure a huge central branch emerges, usually at least as big as the mother culm and this loops higher to continue the climbing operation. The leaves are a very dark, rich green, even in the winter unless exposed to high winds, and a shallow sulcus above the main branch nodes can be seen on the larger culms.

In an oceanic climate C. *valdiviensis* is extremely wind tolerant but this is most unlikely to be so if they are drying winds. It is only just frost hardy so that, even in mild areas, it is repeatedly checked by the winter, which is probably fortunate given its reputation. If you can grow *Gunnera manicata*, you should try this species. Although if unprotected it would die at the first cold spell, a little cover to prevent the roots and lower parts of the culms from freezing will enable it to regenerate each year into an interesting, but small, plant. In many parts of Europe it has been confused with C. *ramosissima* and is still occasionally seen under that misnomer.

DREPANOSTACHYUM

A genus of Himalayan bamboos from dry sub-tropical forests. They will withstand a few degrees of frost and should be given a sheltered mild position with good air humidity, and, in areas other than the more northern latitudes, some shade as well. All members of the genus are delicate in appearance, with a very compact root system. They usually have very small, thin leaves and multiple branches of uniform size and almost horizontal orientation, often giving a 'pompon' profile. This is seen particularly on older culms. About 25 branches are formed in the first year, increasing to up to 70.

Without close inspection, members of this genus can be distinguished from most other genera with similar cylindrical culms by their delicate stature. The only genus with which they can be confused is *Himalayacalamus*, which has fewer branches that vary in size and are more upright. The culm sheaths are rough inside and narrow at the apex. The genus is generally less hardy than *Himalayacalamus* but endures drier conditions.

Drepanostachyum falcatum

Min. -4°C (25°F), Culms 6m x 2.5cm (20ft x 1in), Leaves 10cm x 6mm (4 x ¼in). This is one of the most elegant and easily recognized bamboos, if it is grown in the cool, protected and moist conditions that it enjoys. It was introduced into Britain by someone called Nees in about 1870 and is short-lived (about 15 years between flowering) but can be very quick growing. A well-grown plant can have slender culms that arch gracefully under the weight of luxuriant foliage until they touch the ground. Its leaf formation is particularly attractive, pointing downwards in regular sprays. The leaves are small but linear and narrow, except for terminal leaves, which can be much larger, and they are a uniform mid-green if given some shade. The culms are also distinctive, small in diameter compared to the length and a uniform grey-green except for a small purple staining just below the nodes. It is one of the least hardy bamboos, often losing all top growth, but if the root is protected from freezing it will regrow each year, provided the summer conditions are to its liking. The resulting plant will probably then be no more than 2m (6ft) tall, but it will be very attractive.

This species has often been confused with *Himalayacalamus falconeri*, which has much more extensive purple staining at the node. They are distinctly different when both are growing well. A study of the distinguishing features of both genera helps identification.

D. falcatum is tolerant of hotter conditions with drier air but will grow very differently, becoming more rigid and upright with yellow culms and yellow-green leaves. The leaves also lose their distinctive downward-pointing formation. Because of this tolerance, in conjunction with its ability to withstand drier soil and its very small root system, it makes an excellent pot plant for a light spot in an unheated room.

Drepanostachyum intermedium

Min. -9°C (15°F)?, Culms 4m x 2cm (13ft x ¾in), Leaves 13 x 1cm (5 x ½in). This species has a more open root structure than most of the genus. It is identified by its persistent leaf sheaths with oral setae and large auricles. The culms also have swollen nodes and the leaves are mid-green. It is not common in cultivation but can be obtained from one or two specialist growers.

Drepanostachyum khasianum

Min. -9°C (15°F), Culms 5m x 5cm (16ft x 2in), Leaves 5 x 1cm (2 x ½in). A compact upright species with small leaves. The culm sheaths have no auricles or oral setae and the culms are large and glossy mid-green with a crimson stain above the node.

A good proportion of the new culms are killed or partly damaged by the winter cold and it needs a regular manicure to keep it looking good. Nevertheless, it is a very elegant and stately plant that is ideal as a specimen in mild, sheltered gardens.

The plants in cultivation named as this species are thought by some authorities to be the same as *Himalayacalamus falconeri* but with me it seems distinct, with more branches and smaller leaves. It has just flowered, however, and it will be a number of years before the seedlings are old enough to study.

Drepanostachyum microphyllum

Min. -9°C (15°F), Culms 4m x 4cm (13ft x 1½in), Leaves 5 x 1cm (2 x ½in). Very similar to *D. khasianum*, but

Drepanostachyum intermedium needs a sheltered position with slight shade and wind protection from surrounding plants.

with even smaller leaves and culms that mature to purple, this is also a very desirable species. In my garden it is not damaged quite so much as *D. khasianum*, although this could be due to local conditions. Some authorities regard the plants in cultivation under this name to be *Himalayacalamus asper*.

Fargesia

A large and very useful genus, found between 1000–4000m (3300–13,100ft) or occasionally higher in the Himalayas. It incorporates some of the most hardy species in cultivation, but also some that can only be grown in favoured locations. Many potentially useful species await introduction and some that are now rare in cultivation may prove to be ideal garden plants. Members of this genus can be very wind tolerant. Although often losing their leaves during the winter gales or if subject to very cold temperatures, they produce a bright, clean flush of new leaves in the spring when most bamboos are looking at their worst.

The genus is primarily distinguished from others with cylindrical culms by its branch formation. Most members have about 4–5 small branches at each node,

A recently planted specimen of *Fargesia albocerea* enjoys this selected corner, sending out lush growth.

although some species or some very old culms may have up to 15. They extend about halfway around the node and have a characteristic arrangement – sloping outwards from the culm at a 45-degree angle. On most cultivated species, they are short and do not re-branch. The roots are pachymorph and are usually compact, but in heavy shade they can open out so that the culms are spaced several centimetres apart. Leaves and growth are not coarse but are generally less delicate than in *Drepanostachyum* and *Himalayacalamus*.

A number of new species are now being introduced into the Netherlands by Jos Van de Palen and other growers, and some of these have great horticultural potential. The most useful of these are included in the descriptions below for, although they are currently difficult to obtain, they will almost certainly become more widely known within a few years. As these new plants are received it has become apparent that many of *Fargesia* species, including some of those that have had one clone in cultivation for many years, have many quite distinct naturally occurring forms. This is also found in that other mountain genus *Thamnocalamus*.

The genus is at present under taxonomic review. Some members have recently been transfered to the genus *Borinda* by Chris Stapleton who has extensively studied these and other Himalayan genera (see Reading about Bamboos, p.154) but this genus has yet to be widely recognized. *Borinda* was created five years ago to accommodate those species that have little horticultural differences from the true *Fargesia* species but have flowers similar to the yushanias (but without the extended rhizome system). Horticulturally, borindas differ from fargesias in having deciduous leaves if hardy, or persistent leaves if less hardy. They originate from Nepal through Yunnan to Vietnam, while *Fargesia* species are found in Central China in Gansu, Qinghai and East Sichuan.

Fargesia albocerea (*Borinda albocerea*)
Min. -10°C (14°F)?, Culms 4m x 2cm (13ft x ¾in), Leaves 6 x 1cm (2½ x ½in). This is a very new introduction of unproven hardiness. The culms are initially covered in a blue-grey bloom but later age to yellow. Its other distinctive features are its small number (1–5) of very long branches and its soft wide leaves that have a bluish sheen. Several different clones have been introduced so there are wide variations in form and hardiness.

Fargesia angustissima (*Borinda angustissima*)
Min. -9°C (15°F), Culms 7m x 2cm (24ft x ¾in), Leaves 5cm x 5mm (2 x ¼in). This is a large but not very hardy species presently grown in the USA. It has heavily white-bloomed new culms with longitudinal striations. The culm sheaths are initially purple with no auricles, and the narrow blade is sharply reflexed so that it

points downwards. There are 5–10 branches bearing distinctive narrow leaves.

Fargesia denudata

Min. -23°C (-10°F), Culms 5m × 1.3cm (16ft × ½in), Leaves 5 × 1cm (2 × ½in). A recent introduction with great potential and charm, making a fine specimen plant for cooler regions. The leaves and culms are mid-green with a tinge of yellow but the latter can become clear rich yellow in a sunny location. Masses of small, wide leaves produce an elegant drooping profile that can be confused with a well-grown *F. murieliae*. However, it is clearly distinguished by its shorter leaves, its large quantity of small, very short branches, its overall yellow effect and its leaves, in sprays of about four, that are held parallel to their supporting twigs. (The leaves are often approximately at right angles in *F. murieliae*.)

This description relates to the introduction by Roy Lancaster that was made a few years ago: there are different forms of this species now being introduced.

Fargesia dracocephela

Min. -23°C (-10°F)?, Culms 5m × 1.3cm (16ft × ½in), Leaves 10 × 1.3cm (4 × ½in). Another leafy, very hardy, recent introduction with a similar profile to *F. murieliae*, this species grows best in a sunny location, when it produces a wealth of mid-green leaves and yellow-green culms. It is distinguished from other members of the genus by usually being more upright with layered leaves, and is clearly identified by its leaves which gradually taper to a fine point at both ends (most bamboos have leaves with rounded bases). It also differs from *F. murieliae* as its leaves are held approximately parallel to the supporting twigs in bunches of about three. If leaves of both are available for a direct comparison, *F. dracocephela* will be found to have relatively hard leaves.

The many plants in cultivation all arise from seed distributed a few years ago by Max Riedelsheimer, and in time various distinctive clones will probable emerge.

Fargesia frigidorum (*Borinda frigidorum*)

Min. -23°C (-10°F)?, Culms 3.5m × 1.7cm (11.5ft × ½in), Leaves 2.5cm × 5mm (1 × ¼in). As suggested by its name this species is reputed to withstand considerable cold and could be found to be at least as hardy as *F. nitida*, although it is completely deciduous in the winter. This plant is not available at the time of writing but has considerable potential particularly in the colder regions. It is a delicate-looking plant with tiny leaves, yellowish culms and elegant form.

Fargesia fungosa (*Borinda fungosa*)

Min. -10°C (14°F)?, Culms 4.5m × 1.5cm (15ft × ½in), Leaves 10 × 1.5cm (4 × ½in). This species has recently been sent as seed to North America from Yunnan, and was initially rejected in many cooler regions as being not very hardy. However, judging by conditions at its collection site, it should be at least as hardy as *F. dracocephela* and a large plant has survived several winters in Kew Gardens, a location that is too cold for most species with borderline hardiness. Seedlings are much less hardy than a plant that has developed woody culms, so it would seem that we have been too dismissive of this species.

The culms are mid-green but can turn red in locations with strong light. New shoots are purple-red and covered in black hairs. The culm sheaths are persistent and covered in dense black hairs. Together with the comparatively large leaves, these hairs are an important identification feature. The leaves are hard in texture with hairs on the underside. The fact that the leaves are carried in bunches of 3–4 could be distinctive, although I am told that this feature is common to all borindas that are not deciduous.

Fargesia murieliae

Min. -29°C (-20°F), Culms 4m × 1.3cm (13ft × ½in), Leaves 8 × 1cm (3 × ½in). Since its introduction into the USA in 1907 and into Europe in 1913, this species has fully deserved its rapid spread. It is very hardy and is now probably the most extensively grown species in the cooler regions of the West, being popular in Scandinavia and doing well in the colder parts of Eastern Europe and the North American continent.

Several good cultivated forms were selected, but these have now been lost during the recent flowering phase that started in most areas in 1978. There have been no reports of any plants surviving full flowering. A multitude of new forms has been discovered among the vast number of seedlings that have been raised, but these have yet to prove their consistency and usefulness in the garden. A form SABE 939 with minor botanical distinctions but probably greater cold

This *Fargesia denudata* was planted behind a flowering *F. murieliae* and has now taken its place beside an *F. nitida* (left).

tolerance was collected recently but this flowered before reaching maturity. Seedlings are being raised and so it could be seen in the future.

The culms are light green fading to yellow-green, and arch gracefully outwards under the weight of the mass of soft mid-green leaves, producing an elegant plant for a prime location. The culm sheaths have a rounded apex with no auricle or oral setae and the ligule is short. The species is readily distinguished from other similar species by its overall soft green colour and by its leaves that, on the more upright culms, are generally set at right angles to the small twigs that support them. The leaves are orientated to suit the light so this positioning can vary but is usually very apparent.

Regrettably the old generation of forms is now lost, including the variegated 'Leda' and the vigorous 'Weinhenstephan'. Some of the first plants to flower did so during 1976–79 at Thyme's Nursery, Koge, Denmark, and selected seedlings from a dwarf form can now be found. These include **'Harewood'**, growing about 1m (3ft) high, **'Simba'**, which has close-packed culms 2m (6ft) high, and **'Thyme'**, which is similar but 1.5m (5ft) high. Other normal-sized forms are being marketed. **'Jumbo'** which has wider leaves than the type, is stronger growing and is claimed to grow taller.

Fargesia nitida

Min. -29°C (-20°F), Culms 4m x 1.3cm (13ft x ½in), Leaves 5 x 1cm (2 x ½in). This is another fine early introduction, with seed reaching Europe towards the end of the nineteenth century. A very hardy plant for cold areas, its elegant form and delicate features belie its tough constitution for it makes a fine windbreak. If severely stressed, however, it will lose all its leaves, and its good looks, until spring.

The culms are similar in size to *F. murieliae* but in most situations quickly turn purple. It is usually a more upright plant with leaves smaller, but also more narrow than *F. murieliae*. They are generally in bunches of about two and set at right angles to the twigs on the upright culms. The culm sheaths have no auricles or oral setae and the sheath is tapered at the apex. The leaves do not appear until the second year on developing culms, and the bare current year's upright culms can usually be seen rising above the older culms, which bend over in varying degrees depending upon the luxuriance of the foliage.

In very heavy shade the purple colour of the culms does not develop and the blue-grey bloom found on the young culms persists. The leaves are also smaller. This is a variable species not only between the many clones selected from the original seedlings but also with growing conditions. It is usually readily recognised by its dark culms, which are upright when new and its very small, narrow leaves.

Many selected forms are popular in colder regions where the range of other *Fargesia* species that can be grown is limited. The forms fall roughly into three groups: forms with smaller leaves than the type and culms that curve at the base include the prolific **'de Belder'** and the slower-growing **'Eisenach'**; forms with upright culms and more normal leaves include the narrow-leaved **'Chennevieres'** and **'Ems River'** which has more consistently purple culms; and forms with longer and narrower leaves with usually a more lax weeping style include the sun-hater **'McClure'** and the very elegant **'Nymphenburg'**.

'Anceps' (not to be confused with *Yushania anceps*) is very similar to 'Nymphenburg' but withstands summer heat much more than any other form. In addition it is more open at the roots. These properties make it very popular with the nursery trade, particularly in warmer zones.

A form of *Fargesia robusta* can have bright red node and branch bases for most of the year.

Fargesia papyrifera (*Borinda papyrifera*)

Min. -10°C (14°F)?, Culms 8m x 6cm (26ft x 2½in), Leaves 10 x 1.2cm (4 x ½in). This is another, as yet unavailable, species with potential as it combines some hardiness with very attractive large culms, although it is probably less hardy than originally thought. It has an open habit with upright, dull yellow-green culms that age to ochre. There are 3–7 branches per node and the leaves are large.

Fargesia perlonga (*Borinda perlonga*)

Min. -15°C (5°F), Culms 5m x 2cm (16ft x ¾in), Leaves 8 x 1cm (3 x ½in). Recently introduced from Yunnan, this species is proving to be reasonably hardy. In appearance it is quite similar to *F. fungosa* but has tiny or absent auricles and fewer oral setae. The culms are more colourful, nearly solid and have persistent sheaths. It normally produces shoots in the autumn when growing well and in a climate similar to that of its natural habitat.

Fargesia robusta

Min. -15°C (5°F)?, Culms 4m x 1.3cm (13ft x ½in), Leaves 13 x 1.8cm (5 x ¾in). This species is a comparatively recent introduction but has proved to possess very good gardening qualities. It is easy to identify as, unlike most cultivated *Fargesia* species, it has an upright stature.

Fargesia rufa is a recent introduction with great potential. Its small stature makes it suitable for growing in small and courtyard gardens.

The culms are dark green with short branches that grow in a typical *Fargesia* 45-degree angle from the culm, together forming a cone shape. The sheaths are a startling clear white, another identifying feature, and are outstanding in the spring when the new shoots rise from the ground. The geometric pattern of the regular white culm sheaths against the fresh green background of the expanding culms is a special event on the calendar – this species should always be planted in a location where this can be seen.

There is little need to study botanic details to name this species as it has several clear features. However, there are several forms in cultivation, two of which are distinct and desirable and, although rare, these may cause confusion. One has much larger leaves and is more robust and is the only form that justifies its name. The other has a remarkable all-year red and green contrast at the branch point.

Fargesia rufa

Min. -15°C (5°F)?, Culms 3.5m x 1cm (11ft x ½in), Leaves 8cm x 7mm (3 x ¼in). This very new introduction is listed by some sources as being one of the most cold resistant, but as it originates from relatively low altitudes the estimate given is probably more realistic. The leaves are linear and it has 4–13 branches per node, arranged in the typical cone shape.

Fargesia utilis

Min. -15°C (5°F)?, Culms 5m x 2.5cm (16ft x 1in), Leaves 9 x 1.5cm (4 x ½in). For a *Fargesia* this is a coarse plant. It has small delicate leaves but otherwise it is large in all its features. It is fast growing and difficult to accommodate in a small garden. The culms arise at a shallow angle from the compact root system before turning upwards and, consequently, it takes up a lot of room and should not be placed close to a path or building. In addition to this vase-shaped base, it has a tendency for the outer culms to arch under the weight of the foliage. I have seen it very effectively placed in a large garden on the banks of a lake.

The sheaths have no auricles or oral setae. It produces 7–18 branches, which can re-branch on occasions and are relatively robust. These features can cause confusion with the genus *Thamnocalamus*, although they are only to be seen on occasional nodes or culms, whereas sturdy branches with their re-branching characteristics are the norm with *Thamnocalamus*.

Fargesia yulongshanensis Hort (*Borinda* sp.)

Min. -29°C (-20°F)?, Culms 7m x 3cm (23ft x 1¼in), Leaves 6cm x 5mm (2½ x ¼in). A very rare recent introduction, listed here because of its garden potential. Most species in cultivation are not the true species but their name is unknown. The true species (whose dimensions and hardiness are given here) is found up to 4200m (13,800ft) above sea level in conifer woods in Yunnan, and is the highest Sino-Himalayan bamboo. It is potentially our most cold-resistant bamboo and this, coupled with its large size and delicate appearance, suggests a great future. Its new culms are covered in a blue bloom and it can be identified by the ring of dark hairs at the node. The true species is known to be growing in at least one Western location but the other plants in cultivation under this name are proving to be fairly hardy, too. The immature plants of the latter are an overall rich green with small diameter, elegantly arching culms and small quantities of narrow leaves. It seems to be related to *F. frigidorum*. Several similar species are currently under observation to find one that combines elegance and hardiness.

× Hibanobambusa (× Phyllosasa)

This genus comprises only one species plus one form. The type plant is a generic cross that was found in the wild after the simultaneous flowering of *Sasa veitchii* and *Phyllostachys nigra* 'Henonis' in the same locality, and is possibly, therefore, from those parents. It shows the characteristics of both genera in a very effective way and the garden qualities of both the species and the form are unique.

× Hibanobambusa tranquillans

Min. -20°C (-4°F), Culms 5m x 2cm (16ft x ¾in), Leaves 23 x 4.5cm (9 x 1¾in). This has combined the large culms and the prominent nodes of *Phyllostachys* with the large leaves of *Sasa*. The culms and branches are similar to *Phyllostachys* with grooved internodes and usually two robust branches per node but these features are proportionally more slender. The leaves are more reminiscent of *Sasa palmata* than *Sasa veitchii*.

As one would expect, the rhizomes are leptomorph but fortunately not quite as rampant as *Sasa*, although in time it will certainly need some control. It produces a ball-shaped bush of dense dark green foliage about 5m (16ft) diameter that hardly shows any damage even after the worst of winters.

The form **'Shiroshima'** has some of the largest and most brightly variegated evergreen leaves available to the gardener. They are very clearly and strongly striped in conspicuous creamy-white, which can often show pink or purple shades when young or in strong sunlight. It grows as large and as strongly as the unvariegated plant.

Himalayacalamus

A clump-forming pachymorph genus from cool broadleaf Himalayan forests. Similar to *Drepanostachyum*, the plants are larger and generally more able to withstand the cold but are less tolerant of a dry location. The species in cultivation are, however, not very hardy and only withstand a few degrees of frost. The leaves are delicate and without cross veins (tessellation). This indicates that *Himalayacalamus* species require a location with warm winters and shelter from strong winds.

This very elegant unnamed species is presently known under the incorrect name of *Fargesia yulongshanensis*.

The genus differs from *Drepanostachyum* in a number of ways. The most obvious is the smaller number of new branches, usually about 15 in the first year. Branches also vary in size, are held closer to the culm and do not extend so far around the circumference. The culm sheaths are usually broad at the top with a short ligule. They are smooth inside, whereas those of *Drepanostachyum* are rough at the top. The basal culm internodes increase progressively in length from the base.

Himalayacalamus falconeri

Min. -8°C (18°F), Culms 7m x 8cm (23ft x 3in), Leaves 9 x 1cm (3½ x ½in). This species is a native to north-east Himalaya, Nepal and Bhutan. Large quantities of seed

PLATE X

Bamboo leaf shapes

Indocalamus tessellatus

Sasa veitchii 'Nana'

Fargesia nitida

Thamnocalamus crassinodus 'Lang Tang'

Pleioblastus pygmaeus
Drepanostachyum falcatum
Shibataea chinensis
Chusquea cumingii
Pseudosasa owatarii
Indocalamus hamadae
Scale approximately half lifesize

were sent to Kew in 1847 and were then distributed to other European countries. It is very elegant, usually growing to about half the above size in cultivation. As a large plant it is upright and stately with a very compact root system in the style of the tropical clumping bamboos. As a smaller plant it is beautiful and elegant, with the outer culms arching gently under the weight of the mass of small leaves. It is often confused with that other elegant species *Drepanostachyum falcatum*, which is easily distinguished by its very narrow and uniform rich green leaves and, in good growing conditions, its much more arching culms.

A distinctive feature of *H. falconeri* is the thick jelly-like covering on the new shoots, which is probably there for protection from insect attack. The species is distinguished from other members of the genus by its complete lack of auricles or oral setae. Flowering is approximately every thirty years and most mature plants in cultivation have flowered and died over the 1990s, but plenty of seed was set to secure the next generation.

When growing well, **'Damarapa'** is a stunning form with cream and green variegated culms. It was only the second bamboo species to be introduced into the West in the 1830s and no doubt it contributed to the misplaced opinion prevailing at that time that most bamboos were not hardy. It can grow as large as the species but is generally more upright in its growth at all stages. The variegated culms are heavily striped with random clear cream panels at sheath fall, but quickly attain a red stain on those areas exposed to bright light. The colours are damaged if exposed to cold or wind, and this is particularly so on the late-emerging culms, which are a feature in most growing conditions.

It differs from the species in that it has never been recorded in flower. Also, the older culms often produce large bunches of branches that are fused at the base and have distorted leaves. These branches are sometimes abnormally long and the distorted leaves are like those that are a precursor to flowering. If the branch clusters are removed in one piece, they can be rooted in a humid greenhouse with a little bottom heat, as can a short section of culm along with the node. Neither the rooting nor the branch clustering have been seen on the type plant.

Until recently 'Damarapa' was thought to be a different species, and was known under the name *H. hookeriana* (or *Arundinaria hookerianus*). This is the true name for a very different species with distinctive blue culms, however. The blue bamboo is also in cultivation and care should be taken when purchasing to be sure that you have the correct plant.

Himalayacalamus hookerianus
Blue bamboo

Min. -10°C (14°F)?, Culm 9m x 3cm (29ft x 1½in), Leaves 11 x 1.5cm (4½ x ½in). This is rare in cultivation and is listed here mainly to assist with the confusion that exists regarding its name. It is called the blue bamboo because of its blue waxy new culms. They age to yellow-green or purple-red in a cold location, and have up to 30 branches when mature. In addition to the culm colour, this species is recognized by its long, narrow culm sheaths with a long ligule. Auricles and oral setae are absent and the narrow blade is reflexed. It can grow up to 13m (40ft) in a large greenhouse or a favoured site.

This species had been growing at Kew and at Edinburgh for many years under an incorrect name until it flowered recently when its true identity became apparent. It died after flowering but seed was distributed widely so many seedlings should exist, although they are juvenile at the time of writing and of unknown hardiness. It is a fine plant that will be limited only by its tenderness.

INDOCALAMUS

This genus was once grouped with *Sasa* and is very liable to be confused with that genus, but not with any other. The characteristics of both genera are their large leaves, their relatively small size and their few large branches. These characteristics are subtly different between the two, however. The leaves of *Sasa* are ovate (boat-shaped), and are not more than four times longer than their width. Those of *Indocalamus* are lanceolate, or more linear, and four to five times longer than the width. *Indocalamus* species tend to be taller but not invariably so. Both genera have single large branches. This is always so with *Sasa*, but *Indocalamus* can have two or more branches on occasions. In addition, the nodes are not swollen on *Indocalamus* but are swollen on *Sasa*.

The genus contains some special garden species, but most are only second-rate, have rampant leptomorph

rhizomes, and are best avoided in most gardens. The following are those species with some garden merit. All have dark green leaves and the added advantage of looking good in the winter.

Indocalamus hamadae

Min. -20°C (-4°F), Culms 5m x 1.5cm (16ft x ½in), Leaves 40 x 8cm (16 x 3in). This species has noble, erect, architectural qualities and probably the largest leaves – sometimes longer than 60cm (24in) – of any temperate bamboo in cultivation. It has up to three branches on the higher nodes. The leaves have a tendency to wither at the terminal point during the winter.

Indocalamus latifolius

Min. -20°C (-4°F), Culms 3m x 1cm (10ft x ½in), Leaves 30 x 5cm (12 x 2in). This is another tall, erect species with great architectural qualities, although smaller in every respect than *I. hamadae*. It can spread at the roots but even so is a very desirable garden plant.

One of the few large-leaved bamboos that stays small and is slow to spread, *Indocalamus tesselatus* is highly recommended.

Indocalamus longiauritus

Min. -20°C (-4°F), Culms 2.5m x 1cm (8ft x ½in), Leaves 30 x 5cm (12 x 2in). A smaller species with leaves similar to *I. latifolius* but with a narrow, pointed end. It can spread excessively at the roots.

Indocalamus solidus

Min. -20°C (-4°F), Culms 3m x 1cm (10ft x ½in), Leaves 20 x 3.5cm (8 x 1½in). This species can be identified by its prominent auricles, its very large oral setae (1.5cm/½in) and its culms ageing to blackish-green. It also has an energetic root system.

Indocalamus tessellatus

Min. -25°C (-13°F), Culms 2m x 5mm (6ft x ¼in), Leaves 40 x 7cm (16 x 3in). Apart from the first two species, this is the only other member of the genus that can be recommended for the average gardener, being quite a special small species that does not spread excessively. It is fairly distinct with its mass of large and long spear-shaped leaves bending the narrow culms into a leafy mound about half the stated height. The leaves can be up to 60cm (24in) long.

It was introduced from China into Britain in 1845 and has only been known to flower partially on two occasions. Some plants are flowering at the time of writing.

OTATEA

A tropical genus which includes one species from Mexico that is sometimes used as a potted house plant or as a specimen outdoors in a very favoured location.

Otatea acuminata 'Aztecorum'

Min. -3°C (27°F), Culms 6m x 3.5cm (20ft x 1½in), Leaves 30cm x 5cm (12 x 2in). This elegant weeping form has well-spaced culms and long linear leaves, but grows quite differently in a pot. The light green leaves completely obscure the culms when growing well. It must have a warm and very light location when grown in any temperate area.

PHYLLOSTACHYS

This genus is one of the major divisions within bamboos. All members of this genus have been found to be very useful to mankind for a multitude of purposes and because of this, cultivation has been traditional and extensive, and special clones selected. Most species within this genus make very good, though large, garden plants. It contains the majority of cultivated bamboo forms, often in amazing bright, clear colour combinations, and they offer unique opportunities to garden designers in all temperate zones. They are statuesque or monumental, rather than elegant, in areas with warm summers, and can cover large areas with large, well-spaced culms. In cooler growing conditions some can behave like clump-forming species with small, arching, elegant culms. There are some very cold-resistant species available including some with good colour combinations.

The genus comes from lowland China, mainly north of the Yangtze valley, and therefore withstands a continental-type climate with cold winters and hot summers. In oceanic regions with cool summers, the running feature of the rhizomes may be slow to develop and it then behaves like a clumping bamboo. *Phyllostachys* will withstand very hot summers but, ideally, should then be provided with ample water and some shade from the heat of the day. Experiments conducted in North America indicate that members of the genus benefit from a spell of cold in the winter and start into growth earlier and more strongly after withstanding normal winter weather. In very warm areas, their normal growth is so vigorous, however, that this lack of cold weather is not noticed.

Phyllostachys is unlikely to be confused with any other genera as it has several instantly recognizable physical features. However, the species within this genus are distinguished by only small botanic variations, so are easily confused: plants often have to languish under the title of 'another green *Phyllostachys*' until summer when new shoots and culm sheaths can be examined. In most cases, the natural origin of the species is not clear, and very often it is not obvious if a species is in fact a species or a cultivated clone. Species have spread and even naturalized far beyond their native China and are common in most of the surrounding countries. Like most bamboos, the *Phyllostachys* vary considerably with growing conditions so physical features, as distinct from botanic features, should be used with caution when identifying a plant. For these reasons the plant descriptions consist mainly of distinguishing botanic features. The new shoots and culm sheaths are most important for identification purposes. The colour of these can be uniform or show a multitude of mixed colours in varying patterns. The presence or otherwise of the ligule, auricles and oral setae are also essential identification features.

The genus itself is identified by the deep sulcus, or groove, always present on the culm internode above a branch node. This can also be seen on the branches and on the energetic leptomorph rhizomes. There are usually two branches per node (one large and one considerably smaller). Rarely there is only one and just occasionally three; the third branch is usually very small and often fails to develop. The number of branches does not increase with the age of the culm and the nodes are usually prominent and finely sculptured. The prominence of these features is reduced in larger culms and usually the lower nodes of large culms do not produce branches, so do not have sulci. However, no other temperate bamboos produce very large culms so, although these smoother culms are harder to identify, any bamboo with large-diameter culms is almost certainly a *Phyllostachys*. Sometimes the genus can be confused with *Semiarundinaria* as the latter can possess a partial sulcus, and can be of similar stature, but a more detailed examination will clear any confusion.

A detailed study of most cultivated members of this genus is found in *Bamboos of the Genus Phyllostachys under Cultivation in the United States* by F. A. McClure.

Dimensions given in this book are maximum for the species and only achieved under ideal conditions. In most gardens the plants will be considerably smaller for most species even if well watered and fertilised.

Phyllostachys angusta

Min. -25°C (-13°F), Culms 7m x 3.5cm (23ft x 1½in), Leaves 15 x 2cm (6 x ¾in). This species is one of the most cold-resistant members of this genus. It has green culms ageing to grey with very pale almost white culm sheaths showing lavender veins and small spots. The ligule is tall and auricles and oral setae are absent. The culm sheath blade is linear and straight and pale green. It is very similar to *P. flexuosa* when both are large but can be distinguished by the culm sheath colour.

Phyllostachys arcana

Min. -20°C (-4°F), Culms 8m x 3.5cm (26ft x 1½in) Leaves 15 x 2cm (6 x ¾in). The culms of this species are green, often sinuous, and with a loose powder covering when first exposed at sheath fall. The culm sheaths are grey or have a slight lavender tint and green veins. A few dark spots can usually be seen. The ligule is sloping and auricles and oral setae are absent. The blade is narrow and wavy and bent back. *P. nuda* is similar but does not have dormant branch buds on the lower nodes, which are present in *P. arcana* and are characteristically covered by the nodal ridges.

'Luteosulcata' has a contrasting yellow sulcus. It grows very much as the type and is a useful garden form.

Phyllostachys atrovaginata (*P. congesta*)

Min. -22°C (-8°F), Culms 7m x 5.5cm (23ft x 2¼in), Leaves 12 x 1.5cm (5 x ½in). The culms are green, ageing to grey. The ligule is covered by a short, triangular blade, and auricles and oral setae are absent. The sheath blade is dark green with shades of wine. It can easily be confused with *P. nigra* 'Henonis' but its culms are much more strongly tapered and the nodes are more pronounced.

It is a very quick-growing species and is useful for landscaping or instant effects in the garden. The culms are relatively large in diameter compared to the height, and stand stiffly upright. They are good for producing a grove effect quickly. The culms emit a slight perfume that is noticeable in an enclosed situation. It has the ability to endure wet conditions because it has evolved air channels within the rhizomes and roots similar to *Arundinaria gigantea*, which is also tolerant of the wet.

Phyllostachys aurea

Min. -20°C (-4°F), Culms 8m x 4cm (26ft x 1½in), Leaves 13 x 2cm (5 x ¾in). This species has stout green culms turning soft yellow in very strong light. The leaves also follow this colour pattern. The culm sheaths are pale green to pale rose-buff. They have very short ligules and no auricles or oral setae. The blades are long and strap-shaped. The lower culm nodes are closely packed and distorted to varying degrees. Larger plants are usually consistent with this distortion and are certainly more reliable and grow better than *P. edulis* 'Heterocycla' in cooler areas. Small plants often do not show this feature but the species can always be identified by the consistent slight swelling below the nodes, which can even be seen on the branch nodes. In cool areas this species remains reliably compact but in hotter regions it can be a determined runner. The culms are always rigidly upright which makes this species particularly suitable for hedges and plantings next to paths or roads.

Most members of the genus do not prosper in heavy shade, but *P. aurea* is particularly sensitive to poor light conditions and will remain a diminutive plant indefinitely in these conditions. In very bright light conditions it develops pale yellow culms and yellow-green leaves, and this probably accounts for its specific name. The species was introduced into Europe in the 1870s and has flowered several times in cultivation.

There are several very good coloured forms which, in combination with the interesting culms and upright growth, make very useful garden plants.

'Variegata' has boldly variegated white and green leaves. It grows smaller than the type and is very useful in gardens as it is one of only a few large variegated bamboos in cultivation.

'Koi' has yellow culms with a green sulcus. The colour combination is much softer than *P. bambusoides* 'Castillonis' and the plant needs slight shade to avoid the green sulcus losing its intensity.

'Flavescens Inversa' has the reverse coloration and needs similar lighting conditions to 'Koi'.

'Holochrysa' is a very good form with reliable all-yellow culms even in a shaded location, but in full sun it develops into a rich yellow that contrasts well with the green leaves.

Phyllostachys aureosulcata

Min. -22°C (-8°F), Culms 8m x 3.5cm (26ft x 1½in), Leaves 18 x 2cm (7 x ¾in). The species has rough, dull green culms and yellow sulci. The culms have a tendency to zigzag on the lower nodes. The culm sheaths are pale olive-green with white-cream streaks. Auricles are well developed but can be completely missing on some lower and upper sheaths. Oral setae are small and there is a large convex ligule. The sheath blades are lance-like, short and reflexed. Colour is usually green-buff-wine with cream bands.

In warm areas this species grows rigidly upright and is suitable for planting next to paths and roads, but in cooler areas it can bend from the base and need some support. It is very widely grown in North America, and although only a relatively recent introduction to Europe, it is now very popular and easily obtained. Probably its only negative quality is that it can become very invasive in warm regions, but in areas that make full use of its cold resistance this is not a problem.

The basic green plant is called **'Alata'** as it was not known in the West until after the plant with the yellow sulcus was classified. **'Spectabilis'** is a fine golden-yellow with a green sulcus, and **'Aureocaulis'** is a richer yellow throughout without any green on the culms. The last two are both strong growers and cold-tolerant; combined with their impressive colours this makes them very useful in colder areas.

'Harbin' has green culms with yellow stripes combined with longitudinal ridges. When growing well it makes a fine plant but I have found it a more reluctant grower. There is also a rare form called **'Harbin Inversa'** with the same longitudinal ridges but on culms that are mainly yellow with a few green stripes. Equally rare is **'Lama Tempel'** which is almost the same as 'Spectabilis', except that its shade of yellow is slightly richer, and has fine, random green stripes. It seems to have a stockier growth style with more leaves, too. The yellow in all these forms turns a temporary red if exposed to sunlight. This red coloration fades after a few months and matures to a darker yellow. Although the coloured forms are not as sharply contrasting as those of *P. bambusoides* group, they grow much better in cooler regions and are highly recommended where *P. bambusoides* forms do not perform well.

Phyllostachys aurea 'Holochrysa' produces soft yellow culms with congested internodes even in a slightly shaded location.

Phyllostachys bambusoides

Min. -15°C (5°F), Culms 22m x 15cm (72ft x 6in), Leaves 15 x 3.5cm (6 x 1½in). This is the giant timber bamboo of the Orient. It is very widely grown and has naturalized freely in many countries outside its native China. In most areas it is a large plant with substantial well-spaced bright green culms. The culm sheaths are green to reddish buff. Auricles and oral serae are obvious and the blade is short, lance-shaped with various bands of buff-wine-cream. This species is easily distinguished from all other *Phyllostachys* by its very long and strong branches, its terminal leaves that are larger than any other common *Phyllostachys*, and the very obvious oral setae accompanying them. Although quite hardy, this species needs warm summers to grow well and other members of this genus should be substituted in cool, temperate regions.

The type plant was introduced to Western cultivation by Du Quilio in 1866 and was well established in the USA by 1889. However, 'Castillonis' was originally introduced into Japan from Korea at the end of the sixteenth or beginning of the seventeenth century (*Nikon Chiku-Fu* p.21). The species had become widely cultivated in the West by the time it flowered in the 1960s and afterwards was quite rare for a time. 'Castillonis' was lost at this stage and had to be reintroduced, but several new mutations were produced after flowering. All forms of this species are very prone to aphid attack. Although this does no lasting harm to strong plants, it is unsightly and should be treated on feature plants, young specimens and the weaker forms.

P. bambusoides has some of the most beautiful coloured forms. These are highly recommended and should be grown in preference to similar forms of other

Phyllostachys bambusiodes 'Kawadana' needs careful siting to fully appreciate its subtly variegated culms and leaves.

species in all areas that are not too cold. All forms have bright, sharply contrasting colours and make a great impact.

'Castillonis' is the most popular form with its outstanding yellow culms and green sulcus. It does not get as large as the type but seems to be stronger growing and more successful in slightly colder regions. Freeman-Mitford (p.140) recorded in 1897 that his 'Castillonis' plant sent up some culms with reverse coloration. This form is now called 'Castillonis Inversa'. Both have slightly variegated leaves.

'Castillonis Variegata' has leaves consistently variegated in white, but this is still not an outstanding feature. I have found the plant slow growing and, in my experience, the normal 'Castillonis' is the better garden plant. **'Castillonis Inversa Variegata'** is a striking plant, heavily variegated in white with new leaves that are brightly shaded with orange. This originated in New Zealand and was brought to Europe by bamboo enthusiast Tony Pike. As you would expect with such a highly variegated plant, it is a more reluctant grower and takes a while to become established.

There are four very good forms with variegated leaves only. **'Sub-variegata'** has fine variegations of light green on a dark green base. **'Albovariegata'** in Europe has good variegated leaves with the orange coloration in the new leaves but the variegation fades as the leaves age. The North American **'Albovariegata'** is very heavily variegated to the point where plant strength is compromised. Leaf coloration is similar to 'Castillonis Inversa Variegata' but probably even bolder. **'Kawadana'** has subtle, fine variegations in a brush-stroke pattern on both the leaves and the culms, although on higher internodes there are often wide bands of rich gold. It needs careful positioning as the colours fade in strong light, reducing its impact.

'Holochrysa' ('Allgold') is a desirable form with rich gold culms. This was recorded by R.A. Young (p.28) as arising spontaneously from 'Castillonis' in the USA and the contrast between its dark green leaves and the rich gold culms is outstanding. It is strong growing in a suitable location.

There are also forms with interesting shapes such as the very slow-growing and rare **'Marliacea'** with congested fibres forming longitudinal ridges and prominent nodes. The sinuous **'Slender Crookstem'** has barely discernible lower nodes with dormant branch buds and the culms do not grow to the diameter of the species. **'Slender White Crookstem'** is similar with a white powder obscuring the green culms. Both are rare in cultivation.

'Tanakae' has normal green culms but with bold, circular, purplish-brown spots, thought to be caused by a virus, developing with age. This form is also not widely grown.

Phyllostachys bissettii

Min. -22°C (-8°F), Culms 7m x 2.3cm (23ft x 1in), Leaves 9 x 1.3cm (3½ x ½in). This species has erect grey-green culms and pale yellow-green culm sheaths with a slight wine stain. Ligules and auricles are present, plus a few oral setae. The sheath blade is narrowly triangular. It can be confused with *P. aureosulcata* 'Elata' but the culms are less pubescent and it lacks the striped culm sheaths.

It is a very strong-growing species that remains a dark fresh green through the hardest of winters. It is not tall but it forms a dense leaf canopy preventing most other plants from growing within its spread. For these reasons it is excellent for hedges and windbreaks, and although its form is not outstanding, it is an invaluable garden species.

Phyllostachys decora (*P. mannii*)

Min. -22°C (-8°F), Culms 7m x 3cm (23ft x 1¼in), Leaves 8 x 1.5cm (3 x ½in). This species is easily identified by its distinctive dark green culm sheaths with their clear white or pale green stripes giving the appearance of variegations. The sheath is widely truncated at the apex, and the blade is long and lance shaped. Culms are mid-green and the ligule, oral setae and auricles are all minute. This bamboo is normally very upright and straight with masses of drooping leaves. It tolerates hot and dry conditions. Together with its elegant form and cold resistance, this makes it a fine and very useful garden plant.

Phyllostachys dulcis

Min. -20°C (-4°F), Culms 12m x 7cm (40ft x 3in), Leaves 12 x 2cm (5 x ¾in). This large, strong-growing species is very useful for creating a grove in cooler regions where other species remain compact. The culms are dull green, sometimes with fine cream stripes, and covered with a loose powder at sheath fall. The sheaths are greenish-cream with fine cream stripes. The ligules are obvious and the auricles are also well developed with crinkled oral setae. The sheath blade is long and very wavy.

This fine species was introduced into the USA in 1908, but only recently distributed to Europe where it is in great demand as the culms are well-spaced and quickly attain large diameters for relatively short heights.

Phyllostachys edulis (*P. heterocycla, P. pubescens*)

Min. -20°C (-4°F), Culms 20m x 18cm (65ft x 7in), Leaves 6 x 1.3cm (2½ x ½in). This is a potentially huge bamboo but is normally seen growing considerably smaller. It is difficult to establish, taking several years to get going, and should be planted as a large division. This species is the famous Moso of the Orient and a very special plant in regions that can supply its basic requirements – warm summers, good soil conditions and good light in northern regions. In cooler areas it is possible to produce fine plants with patience. For most gardeners, however, it is probably better to plant stronger-growing species that will give quicker results, even if they are not quite so stunning. If you wish to try *P. edulis* you should not be put off by cold winters as it is quite hardy. In Japan it grows in all areas except for the hostile far north, and has been found to grow better in a warm site with a light fertile soil. In the warmer regions it grows to a large size with no special attention but in the colder areas it needs heavy fertilizing to do so.

This species can be recognized without looking at botanical details. The new culms are pubescent and the culm sheaths are very dark, spotted and hairy. The

The bright yellow of the culms of *Phyllostachys bambusoides* 'Holochrysa' contrasts outstandingly well with its leaves.

Himalayacalamus Falconeri can make an impressive plant in a mild and sheltered position.

ligules, auricles and oral setae are large. In good light the culms mature to yellow or even bright orange, and the tiny leaves and branches are usually arranged in delicate layers. Its new shoots often emerge very early in the year, sometimes in the middle of the winter if temperatures are not too cold.

There are many forms but only two – **'Bicolor'** and **'Heterocycla'** – are available in cultivation, and they are both much prized and difficult to obtain. They are smaller than the type but this is scarcely relevant in the garden as they can still make very large culms.

'Bicolor' is beautiful with soft yellow culms randomly green-striped.

'Heterocycla' has distorted nodes on some of the larger culms. It is unusual to find these distortions on culms smaller than 3m (10ft) high. The internodes on alternate sides are very reduced in length so that the node scar slopes obliquely in alternate directions.

Phyllostachys elegans

Min. -18°C (0°F), Culms 10m x 5.5cm (33ft x 2¼in), Leaves 8 x 2cm (3 x ¾in). This species is recognized by its dull green culms with tiny brown spots and the very small lance-shaped leaves. The culm sheaths are olive-green, sometimes with a lavender tinge. Ligules, auricles and oral setae are present and the blade is strap-shaped and wavy.

Phyllostachys flexuosa

Min. -20°C (-4°F), Culms 10m x 7cm (33ft x 2¾in), Leaves 12 x 2cm (5 x ¾in). This was one of the early plant introductions to Europe where it is now widely cultivated and is common in some areas. Sinuous, glossy green culms that age to yellow and black in a bright location are a major distinguishing feature. The sheath is green with close, purple veins and small brown spots. There is a large ligule but auricles and oral setae are absent. The sheath blade is long and linear. Leaves are variable but most are strap-shaped and dark glossy green, and this can assist with identification with some specimens. In cool regions this species remains small with elegant arching culms associated with a compact or a running rhizome system. In warmer climates it is usually much taller and upright.

Phyllostachys fulva ?MITFORD

For temperature tolerance and dimensions see *P. flexuosa* and *P. nigra*.

An obscure bamboo supposedly with tawny culms, listed by early writers. The plant described by A. H. Lawson still exists at Pitt White (p.142) but does not match earlier descriptions. Its old metal label ensures that there is no confusion, but it is now in very heavy shade and remains green. Transplanted divisions show that it is a clone of *P. flexuosa* that seems to develop its yellow and black culm colouring earlier. A more likely candidate for the name *fulva* is an unnamed form of *P. nigra* that is in cultivation. This develops the orange-brown culm colour of 'Megurochiku' without the black sulcus, but its colour is more reliable and it could perhaps be called 'tawny'.

Phyllostachys glauca

Min. -18°C (0°F), Culms 11m x 4.5cm (36ft x 1¾in), Leaves 12 x 1.5cm (5 x ½in). The green culms with their heavy white powder at sheath fall give this species its name. The sheaths are green suffused with wine and with a few small spots. The ligule is wide and short, and auricles and oral setae are absent. The sheath blade is lance- or strap-shaped. There is a form **'Yunzhu'** that develops bold dark brown spots on the older culms but this feature is not particularly attractive.

Phyllostachys heteroclada (*P. purpurata*)

Min. -18°C (0°F), Culms 5.5m x 1.3cm (18ft x ½in), Leaves 7.5 x 1.3cm (3 x ½in). A small species with sinuous culms and prominent nodes. The purple spear-shaped sheath blades held close to the green culm are also distinctive. The sheaths are green, stained wine, and auricles, oral setae and ligules are very small or absent. Like *P. atrovaginata* this species has air channels in its roots and rhizomes, and can adapt to growing in very wet conditions.

There is a form **'Solid Stem'** with little horticultural distinction except for having solid lower internodes. It is thought to be much more cold resistant, however, possibly down to below -23°C (-10°F).

Phyllostachys humilis

Min. -25°C (-13°F), Culms 5m x 2cm (16ft x ¾in), Leaves 7 x 1cm (3 x ½in). A small running, sometimes rampant species thought by some to be a form of *P. nigra*. It is distinguished by its very dark new shoots and its purple-green new culms. The culms age to orange-yellow in strong sun. The branches are distinctly horizontal and the sheaths have large auricles. It is one of the more cold-resistant species and, therefore, has a place in gardens that experience cold winters.

Phyllostachys iridescens

Min. -15°C (5°F), Culms 12m x 6.5cm (40ft x 2¾in), Leaves 10 x 1.5cm (4 x ½in). A species that is easily recognized by the colour pattern of the older culms. The mid-green young culms age to yellow-green and then develop vertical stripes of dark green, yellow-green, and sometimes brown-green and even clear yellow. The culm sheaths are pale reddish-brown with brown marks. The sheath blades have a purple margin, and auricles are present with purple-red oral setae. There is some similarity in the old culm coloration to that of *P. violascens* and probably some confusion with the identification of both species.

Phyllostachys makinoi

Min. -18°C (0°F), Culms 18m x 6.5cm (60ft x 2¾in), Leaves 10 x 2cm (4 x ¾in). This is a large, upright species with brown-spotted pale green culms and a heavy coating of powder at sheath fall. The culm sheaths are pale green with dense dusky-brown spots. The ligules are small and horizontal with vertical bristles, and the auricles and oral setae are absent. The long and lance-shaped sheath blades are pale green with a white border. It is similar to *P. viridis*, including the 'pigskin' surface of the culms, and can only be distinguished by its heavier coating of powder and the ligule which is convex on *P. viridis*.

Phyllostachys meyeri

Min. -18°C (0°F), Culms 10m x 4.5cm (33ft x 1¾in), Leaves 13 x 2cm (5 x ¾in). The erect, mid-green culms of this species have green-buff sheaths with brown spots. The ligule is small and convex and there are no auricles or oral setae. The sheath blade is very long and narrow and arches downwards on the new shoots. It is very similar to *P. aurea* except for the absence of the shortened and distorted internodes. The culms are upright and are hard and strong.

Phyllostachys nidularia

Min. -18°C (0°F), Culms 10m x 4cm (33ft x 1½in), Leaves 9 x 1.6cm (3½ x ¾in). This is a species that cannot be wrongly identified in the growing season. The green culms usually have very pronounced flared nodes, on new culms they are conspicuously fringed with hairs. The culm sheaths are olive-green with stripes of white and wine. The prominent sheath blades are integral with the large auricles producing an arrow-head effect. The culms are nearly solid, or in the case of form **'Farcta'** completely solid in the lower internodes.

Phyllostachys nigra

Min. -18°C (0°F), Culms 15m x 7.5cm (50ft x 3in), Leaves 9 x 1.3cm (3½ x ½in). This is a very useful species that produces colour combinations not found on other temperate bamboos. It is very popular as an ornamental species in China and Japan as well as in the West and is believed to be the first ornamental temperate bamboo introduced to the West. The first introductions were made into Britain in 1827 (possibly earlier) and it flowered in 1900–1905 and again in 1930–1935. The black form is recorded in *Nikon Chiku-Fu* (p.21) as occurring naturally on uncultivated moorland as well as in cultivation. The colour is said to come true from seed.

The statistics given relate to the form **'Henonis'**, which is in fact the type plant and has rough green culms. When grown in warm areas the new culms are a striking green/grey colour. Culm sheaths vary between cream-buff and ruddy-buff and have short wavy blades and prominent auricles, oral setae and ligules. The true *P. nigra* is smaller – up to 12m x 5cm (40ft x 2in) in warm zones, less hardy and slower growing, and is the only bamboo with culms ageing to a true solid black. It needs good light to produce this colour but strong light can give rise to white patches on older culms. It is slow growing and compact in cool areas but can run excessively in warm gardens. There are various forms, such as **'Hale'**, which are more reliable and develop the full coloration earlier. A similar form, often sold instead of 'Nigra', is **'Punctata'**. It never produces the complete solid black, and some green or green-brown is always present, but it is larger and stronger growing and, therefore, has its uses. **'Boryana'** is at least as vigorous as 'Henonis' and develops large dark brown spots in the second year. **'Megurochiku'** has culms ageing to light orange-brown with black in the sulcus.

Phyllostachys nuda

Min. -26°C(-15°F), Culms 7.5m x 3cm (25ft x 1¼in), Leaves 10 x 1.3cm (4 x ½in). This species is the most hardy of the genus. It is distinguished by its dull dark green culms which have a bright white ring at each node. Its identity is confirmed by the absence of auricles and oral setae on both culm and leaf sheaths. The ligule is small and truncated. The culm sheath is greyish-wine, the basal ones also blotched with wine. New culms are so dark in colour as to appear black and the leaves are a darker green than most *Phyllostachys*. The sheath blade is short, triangular and crinkled. **'Localis'** has large purple-brown blotches on the lower part of the culm. These are often so dense as to appear to be a solid brown.

Phyllostachys praecox

Min. -18°C (0°F), Culms 10m x 5cm (33ft x 2in), Leaves 10 x 1.2cm (4 x ½in). A little-known species, grown mainly for its edible shoots which are said to be delicious. It has upright, well-spaced culms with short internodes. New culms are heavily white bloomed with purplish nodes and sometimes show light yellow-green longitudinal stripes. The sheaths are greenish-brown or light brown with dense brown spots and are also white bloomed. Auricles and oral setae are not developed and the ligule is hairy and curved. The long spear-like blade is inverted and wavy.

'Notata' is the plant usually seen, although this is still rare in cultivation. It has green culms with random yellow stripes in the sulcus area only. It is not a bold coloration, particularly on small culms. Much more impressive is **'Viridisulcata'**, its bright yellow culms having a contrasting green sulcus plus random thin longitudinal stripes over the rest of the circumference. As far as I know this form is not presently cultivated in the West but it cannot be long before it finds its way into the garden.

Phyllostachys propinqua

Min. -20°C (-4°F), Culms 7.5m x 3cm (25ft x 1¼in), Leaves 11 x 1.3cm (4½ x ½in). The culms are matt mid-green, the culm sheaths olive with green veins and a bronze sheen. The ligule is very small and there are no auricles or oral setae. The sheath blades are green with a white border, linear, narrow and straight. The form **'Bicolor'** has just been introduced to cultivation. It has soft yellow culms, bold green sulci and a few other random stripes. The boundary between the colours is not sharp, but it is an attractive and different colour pattern and it should have good gardening potential. The leaves usually curl up at the edges to form a distinct longitudinal trough, making this species easy to recognize at all stages of growth. Consequently they seem narrower than they are, with a very long point.

Phyllostachys nigra 'Henonis' can make an impressive grove. Here, it has remained elegant and compact.

Phyllostachys rubromarginata

Min. -20°C (-4°F), Culms 9.5m x 3cm (30ft x 1¼in), Leaves 12 x 2.5cm (5 x 1in). New culms with their long, slender internodes are glaucous green ageing to grey or yellow-grey. The sheaths are olive to buff, stained or striped red. Auricles and oral setae are absent and the new ligule is a distinctive dark red. The sheath blades are ribbon-shaped and straight. This species grows tall and upright, even in cool areas.

Phyllostachys sulphurea 'Viridis'
(*P. viridis*, *P. mitis*)

Min. -20°C (-4°F), Culms 15m x 8.5cm (50ft x 3¼in), Leaves 10 x 2cm (4 x ¾in). The large, upright, soft green culms with their 'pigskin' surface and their almost-flush nodes are instantly recognizable, and this species can only be confused with the much rarer *P. makinoi* (p.117). The sheaths are generally a light rose-buff with green veins and brown spotting. Sheath blades are narrow triangular or ribbon-shaped. The ligule is curved and convex, and there are no auricles or oral setae. **'Sulphurea'** ('Robert Young') is a beautiful plant with light green culms ageing to yellow-green in low light levels

A recent introduction, *Phyllostachys propinqua* 'Bicolor' will be in great demand because of its unusual coloration.

or soft clear yellow in warmer regions. There are a few random fine green stripes as a contrast. There is also a form **'Houzeau'** with green culms and a yellow sulcus.

In some situations this species is reluctant to become established, or it produces large culms from a slow-growing rhizome system. If this situation develops it needs staking, to enable it to prosper, and higher temperatures to the roots. With patience it will get going but may take some time, and it is definitely a plant for a very sunny location in most cooler areas.

Phyllostachys violascens

Min. -18°C (0°F), Culms 11m x 7cm (36ft x 3in), Leaves 13 x 2.5cm (5 x 1in). Early records of this plant are quite clear about its description and there are no problems identifying it. Its new culms are deep violet, almost black, unless grown in low light levels when they are green. They age during the first year to a dull brown, and over succeeding years to brown striped yellow-green, then to dark brown-purple stripes on a light brown culm. It is very invasive and its thin-walled culms are liable to break or part from the main rhizome during the winter. As it is the new culms that are its greatest attraction I remove all old culms as soon as the new culms start to open their leaves, and this keeps the plant looking good and reduces its energy. The sheaths are deep purple with a convex ligule and obvious oral setae.

There is an inferior plant, widely cultivated under the same name, that does not develop the early violet colour but otherwise has very similar coloration. I do not possess any further botanic details. Ohrnberger lists the former as the species, and the latter, with reservations, as a form of *P. bambusoides*. The gardener should also be aware that *P. iridescens* is often confused with both forms, having very similar coloration as it ages, although generally more green, ageing to dark green, with light green, brown, and sometimes yellow stripes.

Phyllostachys viridi-glaucescens

Min. -20°C (-4°F), Culms 10m x 5cm (33ft x 2in), Leaves 11 x 1.5cm (4½ x ½in). A common mid-green bamboo that is not easy to identify positively without inspecting the new shoots. In cool regions it is an unimpressive plant but in warm areas it is tall and upright. The pale buff culm sheaths are distinctive with their heavy brown spotting and blotching, and there are prominent ligules, auricles and oral setae. Sheath blades are long and narrow and usually slightly wavy.

The species was first introduced into France in 1846 from where it has spread to most of Europe. It is now probably one of the most widely found species in the old gardens of Europe.

Phyllostachys vivax

Min. -23°C (-10°F), Culms 15m x 9.5cm (50ft x 3¾in), Leaves 13 x 2cm (5 x ¾in). Even in a cool climate this species is fast growing and large. It is most likely to be confused with *P. bambusoides* but is much more vigorous, has drooping foliage and much thinner culms. The green culms are minutely ribbed, the sheaths are cream-buff with many brown spots, the ligules are short, and auricles and oral setae are absent. The culm sheaths are pointed and the blades are narrow triangular and very wavy.

'Aureocaulis' is an outstanding garden plant with rich yellow culms. It has random heavy green stripes on small culms grading to very few stripes on large culms. **'Huangwenzhu'** has green culms and a yellow sulcus and has been known to form spontaneously from 'Aureocaulis'.

This species is the most cold-resistant large-size timber bamboo that is useful for construction purposes. However, its culms are much inferior to *P. bambusoides* being weaker, thinner walled and easily split, particularly in high winds or under the weight of snow. The species and its forms all have great gardening potential, particularly in cooler gardens where the combination of cold resistance, large size and good colours is unique.

In a warm site *Phyllostachys sulphurea* 'Sulphurea' can rapidly develop large culms from a small rootstock.

Pleioblastus

This is a very large and diverse genus including some very invasive species, many that have little horticultural merit, and a number of species that are so similar as to seem identical unless examined in detail. Plants are generally of robust or even coarse appearance. There are some outstanding and useful species and varieties, however, and the following list is a selection. All have vigorous leptomorph rhizomes and need to be introduced into small gardens with great care. Some members of this genus can be used as specimen plants provided the rhizome spread is limited. Others are suitable as smaller spot plants but some form of pruning to top growth and rhizomes is then essential to maintain their impact. Larger species are usually upright and tough, making them ideal for hedges and windbreaks. Small specimens are useful for groundcover, once again, provided that their searching rhizomes are taken into account.

Members of the genus have multiple branches, cylindrical culms and persistent culm sheaths, together with a leptomorph rhizome system: the combination of these feature makes identification easy. The genus can be confused with *Yushania* because of the latter's extended pachymorph rhizome. However, there are

Pleioblastus chino 'Elegantissimus' has a useful combination of delicately variegated, narrow leaves and upright form.

only a few *Yushania* species in cultivation and these can be easily eliminated without the need to study the root systems.

During the last century many forms were introduced into the West whose parentage had been lost and these were classed as species, and remain classed as such to this day.

Pleioblastus akebono

Min. -20°C (-4°F), Culms 30cm x 3.5mm (12 x ⅛in), Leaves 6 x 1cm (2½ x ½in). This is a dwarf form with distinctive variegations in the leaves. New leaves shade gradually from almost green at the base to almost white at the tip. This effect is outstanding in the spring but it fades as the year progresses and is retained longest in a partially shaded position. This is one of the least invasive species of this genus and it can be recommended for any garden setting, although sometimes it is difficult to grow satisfactorily.

Pleioblastus chino

Min. -25°C (-13°F), Culms 4m x 2cm (13ft x ¾in), Leaves 20 x 2cm (8 x ¾in). A medium-sized species introduced into the West from Japan in 1875. It is of little horticultural merit as it is not very good-looking, particularly in the winter months. When tall it can be confused with *P. simonii* but the latter has a distinctive clustered branch formation held close to the culm and is usually larger with narrower leaves. When small it can be confused with *P. humilis*, an equally unimpressive species. The latter is smaller in height and leaf, and its culms turn a dull purple with age. There are some fine forms of this species, some of which look very different to the type plant.

The type plant can be seen covering large areas in older gardens to the exclusion of all other species, but its tough constitution can be used to advantage in cold or woodland situations.

'Aureostriata' has good, but not bright, white variegations on the leaves. These are consistent, even in heavy shade. The plant is smaller than the type, tending to arch, and is slower to spread. The very distinct and aptly called **'Elegantissimus'** (or 'Vaginatus Variegatus') is an attractive, variegated form bearing very narrow leaves with fine white variegations. It can grow over 2m (6ft) tall and can be invasive, but in the right setting it is a fine garden plant. **'Kimmei'** is smaller than the type and has leaves striped with a rich yellow combined with yellow and green striped culms. Unfortunately, the very bold, white variegated form **'Murakamianus'** has been flowering for a number of years at the time of writing and is not readily available. It makes an outstanding specimen in an appropriate location in a shaded spot.

Pleioblastus fortunei (*P. variegatus*)

Min. -25°C (-13°F), Culms 1.5m × 5mm (5ft × ¼in), Leaves 12 × 1.2cm (4½ × ½in). A useful subject with consistently and boldly striped cream and green leaves. The root system is normally compact and does not cause problems in a garden setting. As with most plants of this type it looks better for an annual trimming to keep the height down and the leaves new and fresh looking.

This plant was originally introduced from Japan by Robert Fortune and then sold through Van Houtte of Ghent in about 1860. It was distributed to England and some other countries in about 1876. Some plants died during a flowering bout in the late 1970s and early 1980s but these could have been new plantings or weak specimens as members of this genus do not usually die after flowering. Many seeds were produced but the seedlings were all unvariegated. The occasional flower is still being recorded, usually on potted plants.

Pleioblastus gramineus

Min. -20°C (-4°F), Culms 5m × 2cm (16ft × ¾in), Leaves 22 × 1.2cm (8.5 × ½in). The horticulturist will find this species identical to *P. linearis*. It can be differentiated by its leaves having a slight twist at the point and by the culm sheaths, which are smooth and have oral setae. Although invasive, it is a fine and elegant feature plant, provided control is undertaken. It enjoys and looks appropriate in a shady woodland situation where it does not spread so fast.

Pleioblastus hindsii

Min. -20°C (-4°F), Culms 5m × 3cm (16ft × 1¼in), Leaves 18 × 1.5cm (7 × ½in). Although very similar to *P. simonii*, this species is easily distinguished by its abundant upward-pointing branches, twigs and leaves on the upper nodes. This gives each culm a besom-brush appearance. The culms are dull grey and the leaves are also dull. It is a tough utilitarian species but a large specimen can be impressive in a wild garden. It can spread at the roots but is not usually too aggressive.

Pleioblastus humilis

Min. -25°C (-13°F), Culms 2m × 7mm (6ft × ¼in), Leaves 20 × 2cm (8 × ¾in). This is a second-rate garden plant with an invasive root system, which perhaps has a place in a wild or woodland garden. It has upright culms, usually about half the stated height, and relatively large leaves of mid-green. If unchecked it will soon cover a large area to the exclusion of all other species. It is a native of Japan and was introduced into Britain in the later part of the last century. It flowered in 1964–1967 and again over the last few years, and most healthy plants recovered. The seeds are very large and conspicuous.

'Pumilus' (*P. pumilus*) resembles *P. humilis* in habit, size and leaf shape. It has entirely glabrous leaf blades and some other minor distinctions that disappear when it reaches maturity, and for the horticulturist it is identical to the type plant.

'Gauntletti' is listed by earlier writers as smaller, slower growing and less invasive, and I have seen plants that confirm this. However, when divisions from at least two different plants that showed this growth pattern were transferred to my garden, they quickly became identical to the type. Most writers class this as identical to 'Pumilus'.

Pleioblastus kongosanensis f. aureostriata

Min. -25°C (-13°F), Culms 2m × 8mm (6ft × ⅜in), Leaves 16 × 2cm (6 × ¾in). An upright species usually half the above height. The leaves have the occasional yellow stripe and the rhizomes are moderately invasive. It is not a particularly gardenworthy plant but may be of interest to the collector.

Pleioblastus linearis

Min. -20°C (-4°F), Culms 4.5m × 2cm (15ft × ¾in), Leaves 20 × 1cm (8 × ½in). Both this species and the very similar *P. gramineus* are very elegant plants, which, if carefully located, can make outstanding garden specimens. Although there is little to distinguish the two species, they cannot be confused with any other bamboo. They are immediately recognized by their tall, gently arching profile, and upon closer inspection their long, narrow grass-like leaves. The culms are pale green and the leaves rich mid-green. They both enjoy shade and will prosper in dense woodland. Even in dense shade they can run to some extent at the roots, and some regular root pruning is essential if their elegant form is to be maintained. This material is not good for propagation unless it is severed from the main plant several months in advance of lifting, to enable roots to develop.

P. linearis has straight leaves, culm sheaths without hairs, and poorly developed or no oral setae.

Pleioblastus pumilus (see *P. humilis* 'Pumilus')

Pleioblastus pygmaeus

Min. -30°C (-22°F), Culms 40cm x 2mm (16 x 1/16in), Leaves 5cm x 5mm (2 x 1/4in). This is a very variable and confusing species. Early writers describe a plant that is totally different to the specimens seen today. The former grows to the height given but has leaves that are twice as long and are not of fern-like form. The plant can still be found in old gardens such as Pitt White in Uplyme (p.142).

Plants sold under this name today all have small leaves set in two parallel rows and are much more attractive. They vary from 60cm (2ft) high, which was the plant previously called *P. distichus*, to 10cm (4in), which is the plant now called *P. pygmaeus* 'Pygmaeus'.

All the plants can be dwarfed into tiny specimens for the retail market, so it is only after a year or so that you discover if your dwarf lives up to its claims. It is well worth taking time to ensure that it is the smallest clone before handing over your money, as this is a gem and can be planted anywhere except for the often recommended rockery. The larger clones are rampant. Even in the wild garden they are not satisfactory, as they invariably get mixed with similar-sized grasses of equal vigour. In shade they extend and the fern-leaf effect is more open.

Pleioblastus linearis can be invasive but is controlled by shade in a woodland site. Here its elegant arching form is effectively used to frame a boardwalk vista.

Pleioblastus shibuyanus 'Tsuboi' is an excellent small, variegated bamboo but needs trimming to look its best.

Pleioblastus shibuyanus

Min. -25°C (-13°F), Culms 2m x 7mm (6ft x ¼in), Leaves 15 x 2cm (6 x ¾in). This species is only known in the West by the remarkable variegated form **'Tsuboi'**. This has leaves brightly variegated with broad bands of cream and is one of the best small variegated bamboos. It is compact but can actively run at the edges when well established. Like many small bamboos it looks best when clipped frequently and when the clump diameter is kept in proportion.

Pleioblastus simonii

Min. -25°C (-13°F), Culms 5m x 3cm (16ft x 1¼in), Leaves 20 x 2cm (8 x ¾in). A strong, erect species suitable for windbreaks and utilitarian purposes. It can be identified from most species of similar form by the multiple cluster of branch bases that rise from the node in a congested triangular shape. The form **'Variegatus'** is no more ornamental. Most leaves are normal but the plant has the occasional cluster of very narrow leaves with varying amounts of white stripes. Over the years that I have been observing this form, these abnormal growths have always been associated with a few flowers. The variegations are more pronounced on damaged, pruned or potted plants.

This plant is often superficially confused with *Pseudosasa japonica* but upon closer inspection there are few similarities. The branch formations are totally different and the leaves of *P. simonii* are narrower. Although not the most elegant of bamboos, this tough species is very useful for windbreaks and other situations where it is not centre stage, but it can be rather invasive in good soil conditions.

It was first introduced into France in 1862 by M. Simon and spread to the rest of Europe shortly afterwards. It has flowered on several occasions and the large seeds are readily produced.

Pleioblastus viridistriatus (*P. auricomus*)

Min. -25°C (-13°F), Culms 1.5m x 3mm (5ft x ⅛in), Leaves 17 x 2cm (7 x ¾in). This is one of the first introductions and is still one of the finest yellow-variegated garden plants with its rich gold leaves and green stripes. It is typical of the variegated plants introduced as a species. The culms are purple-green and the leaves have a distinctive soft texture. Although classed as running, the rootstock does not give any problems. It should be planted in a good light and the clump cut to ground level every spring to produce a brilliant colour effect. It does not thrive in a dry location or deep shade. It is frequently seen with just one or two flowers but has never been recorded in a full flowering condition.

Occasionally, culms are produced with all-yellow leaves. These are constant and if separated will produce the form **'Chrysophyllus'**. This is also a useful garden plant but it needs careful siting, with light shade, to give good colour without burning. **'Bracken Hill'** is said to be taller growing but in my garden both this and the normal plant grow to 1.8m (6ft) if left untrimmed – they are not an asset at this height as they lose their good looks and bright foliage.

Pseudosasa

Members of this genus are identified by their persistent culm sheaths, and their single large branches (although there can be up to three), which are often on the upper nodes only. The leaves are proportionally much smaller and the rhizomes less invasive than other single-branch genera (such as *Indocalamus*, *Sasa* and *Sasaella*). The culms are cylindrical and the nodes not prominent, and the rhizomes are leptomorph. The main differences from *Sasa* and *Indocalamus* are in the flower structure, which is of little use to the horticulturist. They are mainly upright species of impressive form and suitable as background planting or as specimens. Although classed as running, this genus does not usually cause problems within the average garden.

Pseudosasa amabilis

Min. -10°C (14°F), Culms 13m x 6cm (43ft x 2½in), Leaves 30 x 3cm (12 x 1¼in). A large, erect species grown extensively in one area of China for the garden cane trade. It is distinctive in that the lower nodes are devoid of branches, and in its very hairy new culms and culm sheath. It needs a warm spot to reach its potential size but then makes a regal specimen.

Pseudosasa japonica

Min. -23°C (-10°F), Culms 4m x 1.5cm (13ft x ½in), Leaves 30 x 3cm (12 x 1¼in). A very useful and tough species, widely grown for its permanent rich green colouring, and its imposing and strong stature. It is also good for windbreaks and hedges. It is easily identified by the underside of its leaf which is marked with two shades of green in a ratio of 1:3. Many bamboos have two shades either side of the midrib but this is the only commonly cultivated bamboo with unequal areas. A Signor Fenzi reported in *Gardeners' Chronicle* in 1872 that 'Japonica is not a plant to be recommended for cultivation being affected by a curious disease which causes its culms always to go into flower instead of growth.' As we have seen this pattern for many years in recent times, we can now understand his concern, but most plants are recovered and looking better for their enforced clearing.

It is a native of Japan and South Korea, and was first introduced into France by von Siebold in 1850. From there it was introduced into the rest of Europe a few years later and it was widely grown by the end of the century. Although it withstands low temperatures, its appearance is quickly damaged by the cold, and it is not popular in very cold regions for this reason. In regions with slightly better winters, however, it makes a very good winter subject. It was reported in *Nikon Chiku-Fu* (p.21) that in the wild it grows well on the banks of rivers and is not harmed by some immersion in water. This is assumed to be temporary immersion or with its roots reaching down to the water level, for it will not stand bog conditions. It grows well in raised areas, however, and would be ideal for stabilizing river banks or reclaiming land.

There is a fine form **'Akebonosuji'** ('Variegata') with bold, bright yellow variegation. This form should be divided frequently to maintain the variegation as older plants have more green leaves. Some of the culms of this form will have leaves shading gradually from green at the base to yellow-white at the tip. This coloured form is reasonably constant and, if the culms are separated, is called **'Akebono'**.

'Tsutsumiana' is interesting but not spectacular. The internodes on the larger culms and rhizomes expand above the node to give a curved, bottle-like effect. Unfortunately, the persistent culm sheaths usually mask this feature and it is only on the very large culms that it is noticeable. This form is not as tall as the species and is inclined to spread by producing large quantities of small growth.

Pseudosasa japonica 'Akebonosuji' with a culm (right) showing 'Akebono' leaves that could make a separate plant.

Pseudosasa owatarii

Min. -23°C (-13°F), Culms 1m x 5mm (3ft x ¼in), Leaves 8 x 1cm (3 x ½in). A very useful ornamental small species that can be clipped to smaller dimensions if needed. It has an even more desirable dwarf form originating from Mt. Miyanouradaya on Yaka Island, Japan. The culms and leaves grow naturally to only about half the above dimensions. This dwarf form is very similar to the dwarf form of *Pleioblastus pygmaeus* but has smaller and more pointed leaves with only 3–4 leaves per twig compared to 5 leaves or more on *Pleioblastus pygmaeus*. The running rootstock of both types does not normally cause problems even in a small garden.

Pseudosasa pleioblastoides

Min. -23°C (-10°F), Culms 4.5m x 5mm (15ft x ¼in), Leaves 25 x 2.2cm (10 x ¾in). A very upright species, superficially similar to *P. japonica*, but usually more upright in form. It has shorter leaves of more narrow, linear proportions and the underside has two shades of green on either side of the centre. Usually there is one branch that quickly re-branches, unlike *P japonica* which only re-branches occasionally. Like *P. japonica*, it is a handsome, wind-tolerant bamboo with a dark green overall effect.

SASA

A very large genus of dwarf bamboos, often with only small botanic variation between species. Most are completely unsuitable for a garden setting, particularly a woodland or wild garden where they can quickly dominate the native plants. The larger species have a bold, jungle-like effect, which is difficult to find in any other temperate plant. They are slow to establish, giving a false impression, so be warned. In the wild they dominate all vegetation and can be found on open high mountains and uplands where they resist the extremes of cold. They hinder the development of trees and obstruct the path of the mountaineers. They can grow in open positions as well as shade.

It is tempting to grow them in a pot where they can look impressive, but even here they need to be watched as their invasive rhizomes employ any means available to escape. I know of one plant growing in a large pot in a small front garden. It stands on stone slabs just inside the brick boundary wall and is the epitome of restrained planting. However, the shoots appearing through the tarmac on the road outside produce a completely different impression!

Nikon Chiku-Fu records that in Japan the spread of *S. palmata* 'Nebulosa' is controlled by digging a trench around the plant and filling it with seaweed. Those with a suitable supply of seaweed and who are also thus inclined could use this method with any other of the genus.

The rhizomes are leptomorph and very invasive. The culms are cylindrical and curve from the base. They are small, with one large diameter branch per node. The culm sheaths are persistent and the leaves are large in comparison to the culm. The genus can be confused with *Indocalamus* (p.108, where the differences are listed). The leaves often wither at the edges in winter.

The following selection contains a few that are desirable and a few that can be seen in old gardens.

Sasa kurilensis

Min. -30°C (-22°F), Culms 2.5m x 5mm (8ft x ¼in), Leaves 20 x 4cm (8 x 1½in). This species' native home is the most northern latitude of any bamboo. It comes from the inhospitable Kuril Isles, north of Japan, and also grows on Sakhalin Island in Russia. A *Phyllostachys* species is also reported as growing in Russia but this is thought to have been introduced, making *Sasa kurilensis* the only bamboo representative naturally growing on that continent.

The leaves of this species are clustered together in distinctive palmate bunches at the tops of the slender culms. They are a glossy mid-green and the culms are yellow-green. The species itself is probably too rampant for most situations but there are a couple of useful forms in cultivation. The dwarf form has foliage similar in size to the species, but of a darker and more constant green. Its culms are also slender and about 1.2m (4ft) high in good conditions. It will also make large drifts of dense vegetation and should be kept under tight control.

There is in addition a very unusual variegated form **'Shimofuri'**, that has numerous very fine white lines on the leaves. It will grow as large as the species and needs a little shade to prevent leaf burn. It can run at the roots, but far less so than most in this genus. As it is a very beautiful plant, most growers are delighted if it starts to spread.

Sasa nipponica

Min. -25°C (-13°F), Culms 40cm x 5mm (16 x ¼in), Leaves 15 x 2cm (6 x ¾in). This small species is not typical of the genus, being much less robust in appearance; it still has the invasive qualities of the genus, however. Leaves are 3–5 per branch, narrow and thin with hairy surfaces, and they have a tendency to wither at the edges during winter. Leaf sheaths have obvious setae.

Sasa palmata 'Nebulosa' (*Sasa cernua* 'Nebulosa')

Min. -30°C (-22°F), Culms 3m x 1cm (10ft x ½in), Leaves 30 x 10cm (12 x 4in). This form was introduced from Japan to England in 1889 and was recorded in flower during the 1960s but there are no records of it dying during the flowering phase. 'Nebulosa' (cloud-like) refers to the brown marks always found on the older culms of this plant. Presumably there exists the type plant without this distinctive feature, but if so it is not to my knowledge in cultivation.

It is so tropical in appearance that when it was first grown at Kew Gardens, it was planted in the Temperate House. It now covers many valleys in England and has even been seen growing 'wild' on a windswept moorland. It is impossible to eradicate once well established without destroying all other wildlife, and it should be planted only by those who can exercise constant control. In addition to it being the most common *Sasa* it can be identified by the extensive brown markings on the older culms. **'Warley Place'** has unreliable and indistinct yellow-variegated leaves.

Sasa tsuboiana

Min. -25°C (-13°F), Culms 1.5m x 5mm (5ft x ¼in), Leaves 25 x 5cm (10 x 2in). This is an erect species with large leaves that are subject to slight withering at the edges during cold weather. Slightly less aggressive than most but still needs care when selecting its planting position.

Sasa veitchii

Min. -25°C (-13°F), Culms 1.5m x 5mm (5ft x ¼in), Leaves 25 x 6cm (10 x 2½in). This is another early introduction that can now be seen extensively in old gardens. As it is a smaller species, it is usually seen growing as a carpet on a woodland floor, very similar to how it could be seen in the wild. The culms are purple-green and the leaves are distinctly wide and boat-shaped. The leaves wither prominently at the margins to a parchment-white as soon as the cold weather comes, and this is attractive and distinctive.

There is a small form (about half the height of the species), which has smaller but more boat-shaped leaves that wither more prominently. This is the form to select and it is sometimes sold as **'Nana'** or **'Minor'**. Some growers doubt its existence, claiming that its form is caused by growing conditions. Others claim that it is a distinct species (*Sasa hayatae*) but whatever label it goes under, in my garden it is a consistent and attractive plant.

It has not been recorded as flowering in the West since its introduction into England from its native Japan in 1880.

SASAELLA

A genus that to the horticulturist is almost identical to *Sasa*. The leaves are generally smaller in proportion to the plant height and narrower than most *Sasa* species. All species in cultivation are small and very invasive.

Sasaella masamuneana

Min. -23°C (-10°F), Culms 1m x 3mm (3ft x ⅛in), Leaves 18 x 5cm (7 x 2in). A rampant species only cultivated for its two forms. **'Albostriata'** has striking, bold white variegations that fade to yellow. It is always a distinctive although very rampant plant. **'Aureostriata'** is not so spreading and has larger leaves variegated with dull yellow stripes that fade as the year progresses. Sometimes the effect is good but often the variegations are poor or absent, and it is usually a shabby plant.

Sasaella ramosa (*Sasa vagans*)

Min. -30°C (-22°F), Culms 1m x 2mm (3ft x ¹⁄₁₆in), Leaves 15 x 2cm (6 x ¾in). Probably the most invasive bamboo in cultivation and not a species to introduce into any garden, however tempting its small size and attractive leaf formation may be. It is one of the old introductions and is identified by its small size and withered leaves in the winter. It is distinguished from the smaller *Sasa* species by its small and narrow leaves. It is most likely to be confused with the old form of *Pleioblastus pygmaeus* (p.124), the one without the leaves in two rows. However, in addition to being much more invasive and taller, it also has a distinctve palmate leaf formation.

This species originates from Japan and was sent to Britain in 1892 under the name *Arundinaria vagans*. It flowered extensively during the 1980s and not only did this not check its relentless growth but it also gave rise to numerous naturally sown seedlings.

Semiarundinaria

A very useful genus considered to be a generic cross between the *Phyllostachys* and the *Pleioblastus*. It possesses the features of both. I have seen no records of any plants in this genus setting seed, although flowering is not uncommon. The genus is recognized by its tall and generally upright stature. There are normally three unequal branches, similar in appearance to those of the *Phyllostachys* but not as robust. These can increase in number up to seven on older culms. A shallow sulcus can sometimes be seen, usually on the upper internodes of large culms. Rhizomes are leptomorph and invasive on some species. Culm sheaths remain attached for a while at the centre of the base. The genus is ideal as a windbreak, hedge or specimen.

The genus is native to Japan and China. The Chinese species have recently been transferred to *Oligostachyum* by some authorities, but if it is a generic cross this is a geographic distinction only: to the horticulturist the two genera are identical. The following selection includes those sometimes listed as *Oligostachyum*.

The Chinese genus *Brachystachyum* differs from *Semiarundinaria* only in its floral details, but no species are listed here as they generally have little horticultural merit. Some names listed below are not recognized by all authorities but the species in question are quite distinct horticulturally and make useful garden subjects.

Semiarundinaria fastuosa

Min. -25°C (-13°F), Culms 8m x 4cm (26ft x 1½in), Leaves 15 x 4cm (6 x 1½in). A stately species and one of the best garden bamboos. It has tall, upright culms, typical of the genus, which can form impressive groves, even in cool regions, without the root system getting out of control. Initially dull green the culms age to purple, particularly in good light. The new culm sheaths, if removed before naturally parting from the culm, have a brilliant, shining, wine colour on the inner surface which is not found on other bamboos and is confirmation of the species in smaller specimens.

When mature, this is a distinctive species and unlikely to be confused with other bamboo. It is only outclassed by its form **'Viridis'**, an even taller plant, growing to 12m (40ft) in good conditions. This has culms of a consistent rich green, not ageing to purple, and leaves that are also rich green and slightly smaller than the type plant.

It was introduced into France in 1892 by M. Latour-Marliac, and from there into Britain in 1895. It flowered in 1935–36, 1957, 1965–70, and again recently. Not all plants flowered extensively on each occasion, but most of those that did recovered over time. There have been no records of seeds being set.

Semiarundinaria fortis

Min. -22°C (-8°F), Culms 8m x 4cm (26ft x 1½in), Leaves 15 x 3cm (6 x 1¼in). This species has dark green culms and leaves. The leaves have a hard texture and a sharp taper at the base. The branches are held close to the culm. This species is further identified by the culms and sheaths being covered in small hairs. Auricles absent but oral setae numerous.

Semiarundinaria kagamiana

Min. -20°C (-4°F), Culms 10m x 4cm (33ft x 1½in), Leaves 15 x 3cm (6 x 1¼in). This is quite a vigorous species with older culms turning purple. It is distinctive with its short branches and its absence of auricles and setae on the culm and leaf sheaths. Although relatively new to cultivation this species is not reluctant to spread and it is now readily available through most specialist growers.

Semiarundinaria lubrica (*Oligostachyum lubricum*)

Min. -22°C (-8°F), Culms 8m x 3.5cm (26ft x 1½in), Leaves 13 x 1.7cm (5 x ¾in). A compact species with dark, rich green leaves and culms. The leaves are hard with obvious setae at their bases and the branches quickly re-branch. It is also distinctive in that the two nodal rings are widely spaced and that often the lower culm nodes are completely devoid of branches. Although not kept by many growers, I have found this a good garden species.

In time *Semiarundinaria fastuosa* can become a very large plant and makes an effective windbreak or visual screen.

Semiarundinaria makinoi

Min. -22°C (-8°F), Culms 10m x 4cm (33ft x 1½in), Leaves 9 x 2cm (3½ x ¾in). This species has many similarities to *S. kagamiana* but the larger culms (those over 1cm/½in diameter) consistently have a short raised feature below the nodes. This is reminiscent of that found on the upper nodes of *Phyllostachys aurea* and it extends for about one culm diameter below the lower nodal ring. This species is inclined to run at the roots and can be readily propagated so although it is a fairly new introduction it should be freely available within a few years.

Semiarundinaria okuboi (*S. villosa*)

Min. -18°C (0°F), Culms 7m x 3cm (23ft x 1¼in), Leaves 15 x 2.5cm (6 x 1in). This unusual and rare species has green culms ageing to yellow-green with leaves clustered at the ends of the branches. Terminal leaves are large and *Sasa*-like.

Semiarundinaria yamadorii

Min. -22°C (-8°F), Culms 7m x 4cm (23ft x 1½in), Leaves 18 x 2.5cm (7 x 1in). According to Ohrnberger this is synonymous with the following species, and it is not listed by most oriental writers. However, the plant widely grown under this name is horticulturally distinct from *S. yashadake*. The weight of the leaves tends to bend the culms of *S. yamadorii* which is unusual for the genus. The root system is quite invasive. The leaves are mid- to yellow-green and the culms age to yellow-green and tend to zig-zag. **'Brimscombe'** has soft yellow leaves with pale green stripes, very reminiscent of *Pleioblastus viridistriatus* but a lot less bold. These turn green with age or if not given good light, and it is not sufficiently noticeable for this form to be anything other than a collector's plant.

Semiarundinaria yashadake

Min. -22°C (-8°F), Culms 8m x 4cm (26ft x 1½in), Leaves 14 x 1.5cm (5½ x ½in). This is a fine species with dense leaves and upright habit. The leaves and culms are dark green. It is similar to *S. fastuosa* but is more leafy, and its culm sheaths are covered in brown hairs and lack the colour on the inner surface.

'Kimmei' is a very special plant producing golden culms with a green stripe above the branch bud. This is one of the few plants outside *Phyllostachys* that has bold yellow culms contrasting with dark green leaves. Leaves of this form are smaller than the type and the branches are held close to the culm. There is some question if it is a form of the same species.

Both the species and the form make excellent garden plants and are available from all the major growers.

SHIBATAEA

This is a charming genus, allied to *Phyllostachys*, and all species are very useful in the garden. It is instantly recognizable by its small size combined with short, and normally wide, leaves, and very short branches. The branches have no more than two internodes, do not

Shibataea kumasaca is one of the few small bamboos that is not invasive but even so it benefits from regular pruning.

re-branch and usually terminate in one leaf, although sometimes two can be seen. New shoots are flat and develop into very small-diameter culms with sulci producing an almost triangular cross-section. The culms tend to zig-zag. Although technically leptomorph, in the garden the rhizomes are so slow speading that it can be considered a clumping genus. *S. kumasaca* is the only species that has been in cultivation for a long time and its requirements could be the basis for growing the other species.

Shibataea chiangshanensis

Min. -20°C (-4°F)?, Culms 50cm × 2mm (20 × 1/16in), Leaves 7 × 2cm (3 × ¾in). Very rare in cultivation, this is a plant with great potential. Its small size and small, ovate leaves are particularly delicate. There are usually three branches per node, each less than 2cm (¾in) long, and terminating in one leaf. The leaves are widest near the base. This is an excellent garden plant that can be planted in many garden situations provided its needs are considered (see *S. kumasaca*).

Shibataea chinensis

Min. -25°C (-13°F), Culms 1m × 3mm (3ft × ⅛in), Leaves 8 × 2cm (3 × ¾in). This species is like *S. kumasaca* but it is smaller and an acid soil is not so essential for good growth. There are 3–6 branches per node, each with a single leaf. The leaf is oval with the widest dimension near the base.

Shibataea kumasaca

Min. -25°C (-13°F), Culms 2m × 7mm (6ft × ¼in), Leaves 7 × 1.7cm (3 × ¾in). With few exceptions this will be the species that you will find in established gardens. Although it can grow quite tall, it is usually seen as a dwarf plant less than 1m (3ft) high. The leaves are dark green, ageing to yellow-green, and the branches are very short. It has one to two ovate leaves per branch. The leaves are damaged by temperatures below about -12°C (10°F) but it easily recovers in early spring with a flush of clean leaves. There is a variegated form **'Aureostriata'** that has bold, creamy variegations. Unfortunately, the variegated culms are rarely produced so it is a collector's plant only.

This is a species that benefits from pruning into a neat ball and this is best done after the new culms have expanded. It is said to enjoy an acid soil, but more important than this is copious water in the growing season. It enjoys a warm, damp summer in cooler regions, or protection from heat in areas with hot, dry summers.

A native of Japan, *S. kumasaca* was introduced into England in 1861. It was widely planted in many parts of Europe by the end of the century, and has been recorded in flower only once, in 1964.

Shibataea lanceifolia

Min. -15°C (5°F)?, Culms 1m × 3mm (3ft × ⅛in), Leaves 10 × 1.2cm (4 × ½in). The distinctive lanceolate leaves of this species instantly identify it, otherwise it is very similar to *S. chinensis*, but less hardy. It is quite rare but is available from a few specialist growers.

SINOBAMBUSA

A sub-tropical genus suitable only for milder gardens. Most species in cultivation are tall and upright and superficially similar to *Semiarundinaria*. They are distinguished by their branches, which are equal in size, their elongated internodes (sometimes 80cm/32in long), their prominent nodes and by the culm sheaths, which are shed early and have hairs at the base. The culms are usually compressed or grooved above the branch bud from the mid-internode down.

Sinobambusa intermedia

Min. -7°C (19°F), Culms 5m × 2cm (18ft × ¾in), Leaves 12 × 2cm (5 × ¾in). An upright species with white-hairy new culms. The culms are hairy below the joint, with nodes 50–60cm (20–24in) apart. The branches are normally in threes but there may be two or only one on the lower nodes. The sheaths are green and hairy with auricles and long oral setae. The sheath blade is green and there are 2–4 leaves per twig.

Sinobambusa tootsik

Min. -7°C (19°F), Culms 7m × 4cm (23ft × 1½in), Leaves 14 × 2.2cm (5½ × 1in). The young culms are dark green with nodes 40–60cm (16–24in) apart. The branches are usually in threes with 5–6 leaves per twig. The leaves are lanceolate and very variable in size. The reddish-brown sheaths have brown hairs and purple-red bases with prominent auricles and oral setae. Although the culms are easily marked, which is unsightly, this is a noble plant and is widely grown in

Among the most brightly variegated plants in the garden, *Sinobambusa tootsik* 'Variegata' is best grown in a warm area.

the warmer parts of Japan because of its aesthetic qualities. In Japan it often has its branches trimmed back to one or two nodes to give a pompon effect. **'Albostriata'** has very bold, white-variegated leaves and is also an impressive plant for those able to grow it.

I have found this species to be particularly sensitive to unsatisfactory conditions in a pot, and although it is often recommended for pot culture, I always plant it out at the first opportunity.

THAMNOCALAMUS

A genus of hardy, clump-forming bamboos, mainly from the high-altitude temperate forests of the Himalayas. They can be confused with the other pachymorph genera but are differentiated from *Drepanostachyum* and *Himalayacalamus* by their fewer branches and the tesselation in their leaves. They are clearly distinguished from *Fargesia* by their fewer branches, which are also longer and more robust. All the species in cultivation have branches that rise vertically for the first centimetre and then open out to about 45 degrees from the culm axis. This initial hugging of the culm by the branches is distinctive. There are usually five branches per node, and these re-branch. (Branches are usually short and usually do not re-branch on most *Fargesia* species.) *Thamnocalamus* species are distinguished from *Yushania* by their fewer, more robust branches and by their close-packed pachymorph rhizomes.

The culms are often waxy and the new culms frequently have a blue-grey bloom. The Himalayan species have many naturally occurring forms and sometimes the distinction between species, subspecies and forms is unclear. The divisions below are based upon horticultural differences. The forms can vary in hardiness but this genus contains some of the most beautiful and useful garden plants for the milder areas.

Thamnocalamus crassinodus

Min. -13°C (8°F), Culms 8m x 2cm (26ft x ¾in), Leaves 6cm x 5mm (2½ x ¼in). This is a very variable plant thought by some authorities to be a natural hybrid between two subspecies of *T. spathiflorus*. All its forms are distinguished by their heavy branching, very small leaves and by the obvious swelling above the culm nodes.

One of the most common forms in cultivation is **'Kew Beauty'** with its upright stature and blue-grey new culms, ageing to brown-red, or even a deep red in good light. The branches also age to red.

'Merlyn' is named after Merlyn Edwards, whom we have to thank for introducing these beautiful crassinodus forms. Its culms and branches are green ageing to yellow-green and it is more leafy than 'Kew Beauty'. There is a dwarf green form with a bowl-shaped profile, which grows to about 2m (6ft) high.

Even more rare in cultivation is **'Langtang'** with green culms and the smallest leaves of all, often half the above size.

At Pitt White (p.142) and a few other gardens, a form with startling blue-grey culms can be seen. For a long time this had the unofficial title of **'Glauca'** to identify it within the horticultural trade. It has now

Thamnocalamus tesselatus is not the most elegant of the genus but this young specimen is very impressive.

been called **'Gosainkund'** which is the name of the sacred lake near to where it was collected. This form is not as hardy as the others and suffers leaf loss and damage to new culms below about -7°C (20°F).

These are some of the most elegant bamboos for cool temperate gardens and can be recommended for most locations that are not too cold and have slight shade. All these forms have been collected from different locations and some variation in hardiness would be expected. In the wild this plant often loses all its leaves at the approach of cold weather, so some forms could be more cold-tolerant than stated by adopting this process in some growing conditions. All forms curl their leaves in strong light, which destroys their beauty, but they quickly open out again as shade returns.

Yushania maculata has beautifully coloured culms and is more cold-resistant than most of the genus.

Thamnocalamus spathiflorus

Min. -13°C (8°F), Culms 10m × 2.2cm (33ft × 1in), Leaves 12 × 1.2cm (5 × ½in). The type plant can easily be recognized by its thin-walled culms ageing from glaucous green to an unusual pinkish-brown in good light. The leaves are soft green and the 2–5 branches per node are a similar colour to the culms. Smaller plants are leafy with culms arching gracefully, but larger plants can cover quite an area and become more upright.

'Aristatus' is similar but with culms ageing to yellow-green. There is also a subspecies *nepalensis* that is distinctive with its upright stature and its wider leaves hanging down in palmate bunches.

It is quite wind-tolerant for such a lush species, provided it is not a drying wind and the plant is growing well.

Thamnocalamus tessellatus

Min. -23°C (-10°F), Culms 7m × 2cm (23ft × ¾in), Leaves 8cm × 8mm (3 × ⅜in). Although this is apparently botanically similar to *T. spathiflorus*, unlike most *Thamnocalamus* species, which come from the Himalayas, it is from South Africa and its rugged appearance contrasts with the delicate form of most of the genus. It is rather a shabby plant but is very hardy and tough, and makes a good windbreak. There are between 5 and 8 branches per node, quickly re-branching to up to 12 per node. The branches are generally found on the upper nodes only. The culms are upright and green, ageing to dull purple. The sheaths are persistent and age from pale maroon to white. Its upright form and the colour of the sheaths can cause confusion with *Chusquea culeou* but the latter is a much finer plant that can be distinguished by its stout, solid culms.

YUSHANIA

This genus has similar branching to *Fargesia* but is distinguished by the rhizomes; although classed as pachymorph, each has an extended neck, normally about 30cm (12in) long, and the wide-spaced culms. The rhizome neck is either solid or tubular without any dividing nodes and it has no roots or buds. These features distinguish it from a leptomorph rhizome, which has close nodes with roots and buds. The *Yushania* rhizome neck can extend to 2m (6ft) or more if environmental conditions dictate. All members of the genus are moderately invasive and normally need some control. They are generally found at lower elevations than *Fargesia* species and are, therefore, mainly less hardy. In the cool oceanic climates that they enjoy, they are reasonably hardy, but in areas with hotter and drier summers they are often considered to have questionable hardiness.

Yushania anceps

Min. -18°C (0°F), Culms 4m x 1.3cm (13ft x ½in), Leaves 10 x 1cm (4 x ½in). It is difficult to determine the influence of growing conditions on some bamboos, and this is one of those species. It can be seen as a relatively short plant with vertical culms, or with tall culms, initially vertical but turning through 180 degrees under the weight of a wealth of tiny leaves, the plumes hanging gracefully like no other plant. This variation could be due to climate, soil conditions or it could be a different clone. With good growing conditions the two types in my garden remain constant. There is a famous and particularly large specimen about 9m (30ft) tall growing at Pitt White (p.142), which is considered a form and is sold as **'Pitt White'**. It is the luxurious type with masses of small leaves, and there are both seedlings and divisions for sale as it regrew after the bout of flowering in the 1980s.

The type plant is easily recognized by its uniform rich green glossy leaves and culms, and its widely spaced culms. 'Pitt White' is distinguished by its size and vigour and by its smaller leaves, which are often 6.5cm (2¾in) long by 1cm (½in) wide. This form was originally confused by A.H. Lawson (p.29) with a very different small species called Y. *nitakayamensis*.

Y. *anceps* is native to India and was first introduced into England in 1865. It has flowered several times in cultivation and sets good seed. The last flowering was mainly between 1980 and 1981 and most plants appeared to die afterwards. Usually the dead roots were removed, but there were several reports of new growth occurring after a few years of dormancy in gardens where the roots were allowed to remain.

Yushania maculata

Min. -18°C (0°F)?, Culms 3.5m x 1.5cm (11½ft x ½in), Leaves 13 x 1.2cm (5 x ½in). An outstanding plant with vertical blue-grey new culms contrasting effectively with the persistent brick-red culm sheaths. The culms age to dark olive-green and the leaves are narrow and deep shiny green. The sheath blade is long and narrow, and new shoots are marbled with brown. As with Y. *anceps* this can spread at the roots and it can need control in most gardens. A relatively new introduction, this striking plant has proved to be wind resistant and could turn out to be more hardy than stated.

Yushania maling

Min. -10°C (14°F), Culms 10m x 4cm (33ft x 1½in), Leaves 10 x 1.2cm (4 x ½in). Superficially this plant is similar to a luxurious Y. *anceps*, but upon inspection it is easily distinguished by its grey-green culms and very rough culm surface. It makes an imposing but spreading specimen for larger sheltered gardens in mild areas.

7
PEOPLE AND THEIR GARDENS

La Bambouseraie

A magnificent, shady, tree-lined avenue leads into the estate of La Bambouseraie at Prafrance. As you enter in the height of summer, the intense heat is mellowed, the shade within is cool and the bird song refreshing and unexpected. Occasional shafts of sunlight fight to reach the ground. They illuminate the dark like spotlights, and allow your adjusting eyes to recognize the way ahead and the giant bamboo forest that surrounds you on all sides. You are entering the world of superlatives: the biggest bamboos in Europe, the largest bamboo nursery in the world, and probably the biggest bamboo park in the Western world.

La Bambouseraie is just north of Nîmes and Montpellier in south-east France. Eugene Mazel, its founder, was born not far from here in 1815, and it was to here that he returned in middle age after amassing a large fortune in the grocery trade in Marseilles. In search of wealth he had travelled to China for the purposes of bringing back spices and mulberry trees: silk was in fashion and in great demand and he wanted mulberry trees to feed the silk worms. These were the early days of trade with the East (pp.25–27) and plant introductions from this region. Mazel returned from his travels with a passion for large bamboos, after seeing them growing in vast forests in their native land. This passion was to change his life and bring about his downfall.

Mazel discovered the sheltered valley surrounding Prafrance in 1850. A slight dip in the mountains in an area of suffocating heat, it is sheltered from winds and its soil is a rich, silty loam. Its only drawback at the time was a lack of water. The river Amous, a tributary of the Gardon d'Anduze was only about 2km away, but this was too placid for his vision so he spent 100,000 gold Francs constructing a 5km canal to bring water direct from the Gardon. He established his estate and planted bamboos and trees, including *Phyllostachys edulis* and *Phyllostachys sulphurea* 'Viridis', *Trachycarpus fortunei* and *Sequoia*. They thrived, perhaps too well: 24 years later he needed so many household staff and gardeners that his debts forced him into receivership. Mazel was heartbroken. His dream was repossessed by the bank and soon after he was found floating in the harbour at Marseilles, though it was never established how he came to be there.

For 20 years, nearly as long as the garden had been in existence, the bankers tried to return the land to agriculture, but by now the huge bamboos with their indestructible rhizomes reigned. Many an enthusiastic farmer became demoralized by broken ploughs, and the culms that were replaced almost as quickly as they were felled. The estate was eventually purchased in 1902 by Gaston Negre, a botanist who had been captivated by this unique place and who lovingly restored the magnificent bamboo garden. Once more, a man with a dream was in control. To this day the estate remains in his family and tended by those who cherish it.

Although this succession ensured no further man-made disasters, it has not guaranteed freedom from problems. In the time of Gaston's son, Maurice Negre, the immense floods of 1958 destroyed many a treasured plant. However, the silt brought new fertility to the soil and with it came regeneration and new strength. More

Fargesia murieliae planted by a small bridge. This species is one of the hardiest and most elegant bamboos in cultivation.

recently, the current caretakers Gaston's grand-daughter Muriel and her husband Yves Crouzet witnessed massive destruction when an unprecedented snowstorm damaged large areas of huge culms, splitting them like straws. Although the clean-up operations must have been immense, it was not long before the efforts were rewarded with the emergence of fresh new culms. Today this unique estate is one of the most famous bamboo gardens in the world, attracting 350,000 visitors a year. Garden lovers, botanists, bamboo fanatics or just amazed tourists all flock to see the biggest bamboo forest in Europe. Here *Phyllostachys edulis* can be seen growing over 25m (80ft) tall, and *Phyllostachys sulphurea* 'Viridis' at 15m (50ft) tall, in forests of well-spaced culms within which good-sized palms are dwarfed. The forests filter and abate the intense heat of summer, and you can stroll through subterranean light along cool winding paths. Along with the bamboos there are also tropical waterlilies, lotus (*Nelumbo* sp.), banana plants, camellias, maples, palms and many more species that enrich the tropical atmosphere. La Bambouseraie has over 150 species of bamboo, including the French National Collection. Many bamboos are also available for purchase. This garden is a Mecca for bamboo enthusiasts worldwide – a visit is essential.

BATSFORD ARBORETUM

While Mazel's garden is famous all over the world, that of Freeman-Mitford in Gloucestershire is little known. Algernon Bertram Freeman-Mitford, later Lord Redesdale, had a flair for languages and was employed in the British foreign diplomatic service. He worked in Russia and China, and was in Japan in the 1860s when it opened its doors to the West, as was Ernest Satow (p.21), a young translator also with the British legation. While Freeman-Mitford admired the culture of China, Satow had a great love for all things Japanese, and they both had a common interest in bamboos, so it is certain that there was some interaction between the two men.

Returning to England during the 1870s, Freeman-Mitford was influential in a number of developments around London as secretary to the Minister of Works. His work included responsibility for the Royal Botanic Gardens at Kew, which he enjoyed, establishing a close relationship with all the successive directors and being influential in its development. He was an accom-

plished botanist and had become an authority on the East and lectured widely on these acquired interests.

Then, to his surprise, he inherited the Batsford estate with its classic Georgian house and large but unspectacular garden, and immediately set about redeveloping both to suit his own taste. The house was demolished and rebuilt and the grounds were redesigned as a wild garden – a development of the landscape style of the previous century. It embraced the multitude of new exotic material that was flooding into the West, and the less formal gardening principles that were the fashion.

The garden Freeman-Mitford developed was not an Oriental garden but an amalgam of his experience and ideas. An artificial stream was contrived as there was no flowing water on the site and this was fundamental to Oriental gardening. He planted trees in groups of three where possible, added a hermit's cave and incorporated bronze statues, including one of Buddha. A rest house, modelled on those provided for travellers in Japan, was constructed near the Buddha statue and this was to be the focus for the bamboos.

Over the years, he collected and planted one of the most important collections of bamboos of the time. His experiences in acquiring and growing bamboos in these pioneer years formed the basis for his book *The Bamboo Garden*, which was published in 1896. As well as bamboos, many fine trees were planted – magnolia, pinus, acers and so on. Vistas were created into the shallow valley below and a romantic but useless dairy, based on a Swiss chalet, was built to enhance a view.

After Mitford's death the estate was sold to the first Lord Dulverton. As with many gardens, the gardens at Batsford were mainly neglected during the years between the two world wars, although a few trees were added. This situation continued until the second Lord Dulverton took over the estate in 1956. He had a passion for trees and built on the already-fine foundations that he inherited. He was intrigued by anything rare or beautiful and developed the gardens into an arboretum of quality. By now Mitford's bamboo collection had been reduced in numbers due to neglect or flowering, but the garden still contained some very interesting specimens. In particular a very fine group of *Fargesia*

Thamnocalamus crassinodus is available in several forms, all of which are among the most desirable garden bamboos with their compact rootstocks, elegant forms and tiny leaves.

nitida in a shaded wood by the ponds and stream dates from these early years. It is one of the finest groups of this species to be seen and is the epitome of elegance.

Upon the death of the second Lord Dulverton in 1992 a charitable trust was set up under the control of the Batsford Foundation to ensure that the arboretum could continue to exist. It has grown in importance since then. Recent plantings have helped to make it one of the finest woodland gardens in central England. Many bamboos have been added, most of which are now a good size. There are over 60 species and forms, many more than in Mitford's day.

The best time to see the trees and bamboos is midsummer although if you are a photographer the autumn colours are spectacular and the bamboos are probably still looking good. The combination of continuous quality planting from 1886, coupled with its historic interest and the controlled landscaping, make this garden fascinating for all garden lovers, not just bamboo enthusiasts.

Pitt White

The river Lyme in Somerset spends most of its short life running rapidly through a rich, verdant valley. This countryside is steeped in history, which you cannot fail to sense, even if you are not interested in the many Bronze Age relics in the surrounding hills or the world-famous fossil-bearing strata at its mouth. The weather is mild and the vegetation lush: it is easy to see why this area has been populated for thousands of years.

Pitt White straddles the river and, although itself not old, it stands on a site whose history is as old as mankind. The location is damp and humid, and ideal for wool processing, and the old cellars and work rooms can still be seen. It is believed that monks used the location for wool dying, and more recently wool blankets were produced. Pitt White is approached by road down a narrow lane called Tappers Knapp; by the time you have remembered that 'nap' is a word associated with the wool industry and have got on to wondering what a 'tapper' is, you can see the bamboos forcing their way through the hedge, luxuriously enjoying the same damp, humid conditions.

The house and garden are not currently open to the public but the succession of recent owners have all been most welcoming. After asking permission to invade their privacy, you descend from a rustic verandah to a small lawn, and then again to another lawn. Until it flowered and died in the 1960s, this was the site of a fine *Phyllostachys bambusoides* 'Castillonis'. More recently a few other species have been rescued from overgrown parts of the garden and planted here. Obscuring the river that borders the lawn is a huge *Thamnocalamus crassinodus* 'Merlin'. This is the original specimen given to Dr Mutch by Merlyn Edwards and the parent of all the plants grown under this name today.

Walking down the path on the other side of the river you pass a number of bamboo species before you get to a large *Chusquea culeou*. The soil is wet and poorly drained at this point which this species does not enjoy – the only part that looks in good condition is the growth that has been made up the bank behind.

Above the bank are two other specimens worthy of mention. On the left, as you climb steps that are set into the bank, is a very unusual bamboo of elegant form with blue-grey new culms. For a long time this had no name. It has now been identified as another of Merlyn's early introductions and is called *Thamnocalamus crassinodus* 'Gosainkund'. Although it grows well in this site it unfortunately suffers from considerable winter damage. The other specimen is the very famous Pitt White form of *Yushania anceps*, which is regenerating after its recent bout of flowering. Following the path along the rear boundary of the garden you pass by a very large and elegant clump of *Yushania maling* and then through a grove of tall, invading *Chimonobambusa quadrangularis* culms by the well. Here, there are also many other species to be seen, too numerous to mention by name.

Reaching a corner formed by a public path, which is also the county boundary, and Tappers Knapp, you can backtrack and go through a small gate into the road and down the public path and you will soon find an opening into the second part of the garden (which is situated in Dorset).

This extension, which is now completely abandoned and overgrown, once contained a tennis court and the vegetable plot as well as ornamental gardens. Down by the stream is a clump of the old form of *Pleioblastus pygmaeus*, and higher up various large *Pleioblastus* and *Phyllostachys* species. If you have the stamina to fight your way to the end, through the head-high weeds and brambles, you will be rewarded by another large

collection of bamboos. Prominent among these is a very fine and large *Chusquea culeou* and a *Semiarundinaria fastuosa* and *Phyllostachys nigra* 'Boryana'.

Rejoin and continue along the public path and this brings you back to the house. A large section of the valley downstream was once part of the garden. It is now the site of newly built houses. Here you can also see Pitt White Cottage, formerly a tanner's house and more recently occupied by the head gardener.

The Pitt White collection was built up by Dr Nathaniel Mutch during the 1950s and 1960s and for

Chimonobambusa quadrangularis is a distinctive bamboo with an upright form, and dark green culms and leaves.

about 15 years Alexander High Lawson was his head gardener. Together they worked on the book *Bamboos – A Guide to Their Cultivation in Temperate Climates*, but Dr Mutch is not listed as a co-author. How much each contributed is unknown but Dr Mutch was almost certainly more influential than this suggests.

Lawson left Pitt White shortly after the book was published, and held a number of gardening jobs, including working for Shell Petroleum, before retiring to Ireland in 1988. At the time that their book was written, Pitt White contained the most comprehensive collection of temperate bamboo species in Europe and probably also in the world. Although many interesting species are now lost and parts of the garden are abandoned it is still a very interesting place to visit, with many rare and historic bamboo species. If you are observant, you may get a fleeting glimpse of a deer, or you might see a rare plant or bird in addition to some fine bamboos.

If you have difficulty in finding Pitt White, park in Lyme Regis and take the lane up the beautiful river valley for about 2km. Alternatively, park in Uplyme car park by the public house and walk down the river for less than lkm.

THE NATIONAL COLLECTIONS

The National Collections Scheme in Britain is operated by the National Council for the Conservation of Plants and Gardens (NCCPG) for the purposes of recording and conserving rare plant species and forms. The National Collection holders agree to allow anyone to view their garden during reasonable times by prior appointment, or in the case of commercial establishments during normal opening hours (see Where to See Bamboos p.150). Holders endeavour to maintain a comprehensive collection of at least three specimens of each plant and to maintain records of the latest names and the source of origin. Other than obtaining any new introductions, these requirements can all be problematic for bamboo collection holders with such big plants that mainly originate from a single ancient source.

Drysdale Garden Exotics

The general bamboo collection is held by David Crampton at Drysdale Garden Exotics, Fordingbridge in Hampshire. As this is a commercial nursery, the majority of his large collection is to be found in pots, some of which are very large, and is mostly housed in several large greenhouses. The general collection currently stands at over 190 species and forms, and a good proportion are mainly of interest to the collector or botanist as sometimes the variations are minor, unreliable, or the species does not make a good garden plant.

Of more interest to the majority of gardeners is David's demonstration garden, which is to the rear of the nursery. This is a selected collection of bamboos growing in close proximity with each other and other plants and shrubs, in ideal growing conditions. The rate of growth achieved in this part of England, where summers are hot and winters mild, is amazing compared to most of the rest of the country. However, all the plants were given textbook husbandry when the garden was first planted: the soil was initially deep dug and fertilized, strong plants were used and a regime of watering and fertilizing operated so that in the early years they lacked nothing. Within just three years the plants and garden were mature with strong growth everywhere.

Every plant is choice, so it is unfair to focus on just one or two, but my memory retains a picture of a spectacular *Phyllostachys vivax* 'Aureocaulis', a huge *Chusquea gigantea*, and a large and rare *Phyllostachys bambusoides* 'Marliacea'.

The garden is open Wednesday to Friday plus Sunday. It is within 1km of the centre of Fordingbridge. To find it, start from the High Street (B3078) and take the road to Alderholt. 100m past St Mary's church you will see the lane to the nursery on your left.

Bracken Hill

David McClintock is the brave possessor of the *Sasa* collection, and this includes the closely related genera *Sasaella*, *Sasamorpha*, and *Indocalamus*. These genera are not widely grown, because of their rampant tendencies, and are not, therefore, represented fully in any other private collection. Probably they are of more appeal to those with a scientific interest, and the collection is reinforced most admirably in this direction by David's extensive library and herbarium, collected together over many years, and of course his own technical expertise.

David McClintock is well known in the RHS and Kew, and has a deep interest in other plant families, such as the heathers and native wild plants. He has also written a number of books. In addition to the National Collection, his large garden near Sevenoaks in Kent contains a fine selection of other bamboo genera and also many other plants. Particularly well represented is an extensive collection of *Chusquea culeou* forms gathered from all over the country (these are probably now decimated by the recent flowering phase).

Bracken Hill can be found by taking the hidden lane off the A25 at Platt to the south-east of Borough Green just west of Platt Mill. It is not easy to find so remember to ask for precise directions when you telephone for an appointment to visit (p.150).

GWARACKEWENBYGHAN AND BEECROFT

The only other National Collection is that of *Phyllostachys* forms and this is split between Les Cathery and myself. *Phyllostachys* are large plants and therefore it was considered advantageous to grow them on two sites. This enables the requirement for three plants of each form to be satisfied and allows for different growing conditions to be assessed.

Although the two sites are only about 100km apart they experience quite different climates. Les Cathery's south-facing garden, about 1km outside St Buryan, is in one of the mildest growing areas in Britain, and, although this is not necessary for the majority of the genus *Phyllostachys*, it does give potential for growing the *Phyllostachys edulis* forms to a good size. His is a young garden that is open to the full fury of the Atlantic gales but, at the time of writing, the windbreaks are beginning to take effect and Les's collection of tender and unusual plants have adjusted to their growing conditions, and have made new growth and are showing great potential.

In addition to a very large collection of bamboos, including some very rare plants, Les specializes in southern-hemisphere plants and particularly many *Restio* and other largely South African species.

It is interesting to see which bamboos have responded best to these very exposed but humid and mild conditions. Probably the quickest plant to adapt was *Pseudosasa pleioblastoides*; it is now spreading rapidly and forming a solid windbreak. Also good were *Semiarundinaria fastuosa* and *Pseudosasa japonica*. None of the chusqueas had problems with the wind and most of the vigorous phyllostachys are now growing strongly behind the protection afforded by the still-small windbreak plants.

My own collection numbers over 165 bamboo species and forms, with a bias towards mountain species, and is spread over two sites. The tiny garden by the house is used for raising new plants; the main collection is in a separate small field surrounded by a high stone wall. This walled garden is packed mostly with mature and juvenile bamboos and supporting windbreak plants. The north Cornwall coast is open to the cold north wind and has a climate not much different from most of southern England, except that it has a longer growing season, and the occasional cold spell is usually of short duration and does not penetrate the soil. Of all the plants in the walled garden, I consider the collection of unusual mountain species the most interesting.

The house garden, located in the old centre of Wadebridge, is south-facing and based on a tropical theme with plants selected for their foliage impact and colour (see back jacket). Bamboos are not prominent but new acquisitions are planted in the background to bring on before moving to the main area. A major proportion of new plants are grown from seed, and these and weak divisions are raised in a small greenhouse and then transferred to a protected shade area while they gain cold resistance.

The most outstanding plant in the house garden is a huge *Chusquea valdiviensis*, which was originally planted to be herbaceous and provide summer shade, but it has been untroubled by the last two winters and now swamps the top of the garden and part of the property behind.

Neither Les Cathery's garden nor my own is easy to find so it is advisable to get detailed directions at the time of telephoning for an appointment.

APPENDICES

APPENDIX 1
GLOSSARY

Adventitious Occurring in an unusual position.
Aerial Above ground, eg aerial roots.
Angustifolius Narrow leaf.
Argenteus Silver coloured.
Arundinaceous Reed-like.
Auricles Ear-like projections sometimes found either side of the **ligule**.
Axis The main stem, **rhizome**, branch, and so on.
Blind Of a plant without a growing point.
Bloom Waxy deposit, often seen on new culms.
Branch sheath Protective structure at the **node** of branches.
Caespitose (cespitose) Forming tufts, as in **pachymorph** bamboos.
Chimera (chimaera) A plant comprising tissue from two different genetic sources in distinct zones, as in variegated bamboos.
Chino Of China.
Chlorophyll The green medium, comprising four pigments, through which a plant captures energy from light.
Ciliate Fringed with hairs.
Classification The arrangement of plants into related groups, ie Class, Order, Family, Genus, Species.
Clone A group of plants propagated by non-sexual reproduction and, therefore, botanically identical.
Culm A plant stem that is hollow except at the **nodes**, found in plants such as grasses.
Culm sheath Protective structure at the **nodes** of **culms**, showing a similar structure to those on the branches.
Cultivar A cultivated variety. With bamboos, cultivars are often impossible to distinguish from natural variations and, therefore, the word is not often used in relation to them.

In warm regions, plants from the genus *Phyllostachys* can be relied upon to form a good grove with an oriental effect.

Diaphragm A thin membrane forming a partition (between internodes).
Dicotyledon (dicot) A plant bearing two cotyledons (seed leaves).
Distichus Arranged in two rows (leaves or branches).
Edulis Edible.
Fastuosa Stately.
Fibre A thread-like cell or filament.
Flexuosa 1) Pliable, or 2) bent in alternate directions.
Foliage Leaves.
Forma (form) The lowest subdivision of species, used in this book to describe any division below species.
Giganteus Gigantic, larger than the type.
Glabrous Without hairs (not necessarily smooth).
Glaucous Covered with a bluish **bloom**.
Gramineus Grass-like.
Gregarious In a flock or herd, as in the simultaneous flowering of bamboos.
Hardy A plant that will survive outdoors all the year without protection. This varies with location and is not synonymous with 'tough'.
Horticulture The art of plant cultivation, as distinct from botany.
Humidity The amount of water vapour in the air (see also relative humidity).
Humilis Low growing.
Inflorescence The flowering part, a group comprising the complete flowerhead.
Internode The stem between two **nodes**.
Lanceolate Lance-shaped.
Leptomorph A long, thin, running **rhizome** with both rhizome and **culm** buds.
Ligule The projecting rim at the top of the **sheath** and at the bottom of the blade.

Linear Long and narrow with near-parallel sides.
Maculate Spotted or speckled.
Marbled Mottled or splashed with a contrasting colour.
Meristem The growing tissue, distinguished by its power to reproduce. An area where growth is initiated.
Microclimate An artificial climate or a local variation in climate.
Midrib The continuation of the leaf stalk into the leaf.
Monocarpic A species that dies after flowering (except annuals).
Monocotyledon (monocot) A plant that has a single seed leaf, or cotyledon (grasses, lilies, bromeliads, orchids, palms, and so on).
Monopodial Growth continuing indefinitely (as in the branching pattern of **leptomorph rhizomes**).
Morphology The study of plant or animal form.
Nana Small, dwarf.
Nitrogen (symbol N) An element influential in the promotion of foliage.
Node The point where the leaves, branches or flowers are attached. In bamboos this is the point where a rigid **diaphragm** of fibre and **vascular bundles** adds to **culm** rigidity.
Oral setae Bristles sometimes present on the ends of the **auricles**.
Pachymorph The short rhizomes of **caespitose** bamboos that turn upwards into a **culm** and have only **rhizome** buds.
Palmate Of a leaf in the form of a hand, three or more in a radial formation.
Paramenchyma Ground tissue, thin walled and not differentiated into conducting or mechanical tissue.
Persistent Remaining attached after dying, not deciduous (such as culm sheaths).
Phloem The part of the **vascular bundle** that transports nutrients from the leaves to the rest of the plant.
Phosphorus (symbol P) The chemical element that influences root development.
Potassium (symbol K) The chemical element that influences the development of fruit and flowers.
Relative humidity The ratio of water vapour in the air relative to its maximum capacity at that temperature, a measure of the drying effect of the air.
Rhizome A horizontal creeping stem lying on or under the ground, a type of rootstock.
Rudimentary Underdeveloped.
Setae Bristles.
Sheath A tubular or close-fitting supporting membrane.
Silica bodies Silica particles within plant tissue particularly in the epidermis or outer layer of bamboos.
Sinuous Wavy, winding.
Sulcate With parallel grooves or furrows.
Sulcus An internodal groove.
Sympodial A continually branching habit, the pattern of **pachymorph** bamboo rhizomes.
Taxon (Taxa) A biological category or group.
Tesselated Divided into squares (as with the leaf veins of hardy bamboos).
Transpiration Evaporation of water from the leaf or stem surface.
Type The true species, not a form.
Vascular bundles Groups of conducting and other tissue that transport liquids and nutrients within a plant's structure.
Xylem Woody tissue with lignified (woody) walls for conducting liquids from the roots to the leaves.

APPENDIX 2
BAMBOOS FOR SPECIAL PURPOSES

Exceptionally hardy bamboos
Bashania fargesii
Fargesia dracocephela
Fargesia murielae
Fargesia nitida
Indocalamus tessellatus
Phyllostachys angusta
Phyllostachys aureosulcata and forms
Phyllostachys bissettii
Phyllostachys decora
Phyllostachys humilis
Phyllostachys nuda
Phyllostachys parvifolia
Phyllostachys vivax
Pleioblastus chino
Pleioblastus pygmaeus
Pleioblastus shibuyanus 'Tsuboi'
Pleioblastus simonii
Pleioblastus variegatus
Sasa kurilensis
Sasa palmata 'Nebulosa'
Semiarundinaria fastuosa
Semiarundinaria fastuosa 'Viridis'
Shibataea chinensis
Shibataea kumasaca

Bamboos for wet (but not waterlogged) positions
Arundinaria gigantea 'Tecta'
Chimonobambusa quadrangularis
Chimonobambusa tumidissinoda
Chusquea montana 'Nigricans'
Chusquea uliginosa
Phyllostachys atrovaginata (P. congesta)
Phyllostachys heteroclada (P. purpurata)
Pseudosasa japonica
Sasa palmata 'Nebulosa'
Shibataea kumasaca

Bamboos for a site that is occasionally dry
Bashania fargesii
Chusquea culeou
Drepanostachyum species
Pseudosasa japonica
Sasa palmata 'Nebulosa'
Sasa veitchii
Sasaella ramosa

Bamboos for hedging (in height order, shortest first)
Shibataea kumasaca
Pleioblastus chino and forms
Indocalamus latifolius
Indocalamus hamadae
Fargesia (most common species)
Pleioblastus simonii
Pleioblastus hindsii
Phyllostachys aurea and forms
Pseudosasa japonica
Pseudosasa pleioblastoides
Pseudosasa amabilis
Sinobambusa tootsik
Phyllostachys heteroclada (P. purpurata)
Phyllostachys decora
Phyllostachys atrovaginata (P. congesta)
Phyllostachys aureosulcata
Phyllostachys bissettii

Bamboos for forming groves
Bashania fargesii
Chusquea gigantea (C. aff culeou, C. breviglumis)
Phyllostachys aureosulcata and forms
Phyllostachys bambusoides and 'Castillonis' and 'Holochrysa'
Phyllostachys dulcis
Phyllostachys edulis
Phyllostachys heteroclada (P. purpurata)
Phyllostachys nigra 'Henonis' and 'Boryana'
Phyllostachys vivax and forms
Semiarundinaria fastuosa and 'Viridis'

Bamboos for growing indoors
Bambusa multiplex forms
Bambusa ventricosa
Bambusa vulgaris 'Vittata'
Chusquea coronalis
Drepanostachyum falcatum
Otatea acuminata 'Aztecorum'
Pseudosasa japonica

Groundcover bamboos
Indocalamus longiauritus
Indocalamus solidus
Indocalamus tessellatus
Pleioblastus akebono
Pleioblastus pygmaeus
Pleioblastus variegatus
Pleioblastus viridistriatus and 'Chrysophylla'
Pseudosasa owatarii
Sasa kurilensis (dwarf form)
Sasa niponica
Sasa veitchii 'Nana'
Sasaella masamuneana 'Albostriata'
Sasaella ramosa
Shibataea species

Bamboos for deep shade
Chimonobambusa macrophylla f. *intermedia*
Chimonobambusa marmorea
Chimonobambusa quadrangularis
Fargesia nitida
Indocalamus tesselatus
Pleioblastus gramineus
Pleioblastus linearis
Pleioblastus simonii
Pseudosasa japonica
Sasa species

APPENDIX 3
WHERE TO SEE BAMBOOS

National Collections (open by appointment)

Britain

General Bamboo Collection

D. Crampton, Drysdale Garden Exotics (see Where to Buy Bamboos)

Phyllostachys – cultivated forms (joint collection)

L. Cathery, Gwarackewenbyghan, Boskennael, St Buryan, Cornwall, TR19 6DF.
tel. 01736 810096

Phyllostachys – cultivated forms (joint collection)

M. J. Bell, Beecroft, Park Road, Wadebridge, Cornwall, PL27 7NG.
tel. 01208 812892

Sasa, including Sasaella, Sasamorpha and Indocalamus

D. McClintock, TD, WMH, Bracken Hill, Platt, Sevenoaks, Kent, TN15 8HJH.
tel. 01732 884102

France

Hibanobambusa and Phyllostachys

La Bambouseraie de Prafrance, 30140 Générargues, Anduze, Gard

General Gardens open to the Public

The following is a selection of gardens that are freely open to the public, and where bamboos can be viewed. The list is not intended to be comprehensive, and I have been unable to verify all the gardens so it is not an indication of quality.

Canada

Buchart Gardens, Victoria BC
Van Dusen Botanical Gardens, Vancouver BC

France

Arboretum de l'Ecole du Breuil, Paris
Bambous de Planbuisson, Le Buisson
Conservatoire Botanique, Brest, Brittany
Conservatoire des Jardins, Chaumont-sur-Loire
Disneyland, Paris
Jardin Botanique, Strasbourg
Jardin de la Pagoda, Paris
Jardin des Plantes, Paris
Jardin Exotique du Val Rahmen, Menton
Jardin Japonais de Monaco
Jardins Albert Kahn, Boulogne-Billancourt
La Bambouseraie de Prafrance (see National Collections)
Parc Andre Citroen, Paris
Parc Borely, Marseille
Parc de la Pépinière, Nancy
Parc de la Tête d'Or, Lyon
Parc Floral de la Source, Orleans
Parc Oriental de Maulevrier
Phoenix Park Floral de Nice, Nice

Germany

Bodensee, Mainau Island
Botanic Garden and Botanic Museum, Berlin-Dahlem, D-14196 Berlin
Botanic Garden, Nympenburg, Münich
Gruga Park, Essen
New Botanic Garden, Hamburg-Flottbeck
Palmengarten, Frankfurt
University Botanic Garden, Bonn, D-53115 Bonn
University Botanic Garden, Erlangen, D-9105A Erlangen
University Botanic Garden, Karlsruhe, D-76131 Karlsruhe
University Botanic Garden, Oldenburg, D-26121 Oldenburg

Britain

Abbotsbury, Dorchester, Dorset
Arduaine Gardens, Oban, Scotland
Batsford Arboretum, Moreton-in-Marsh, Gloucestershire
Brantwood, Conistonwater, Cumbria
Bridgemere Garden World, Nantwich, Cheshire
Carwinion, Falmouth, Cornwall
Endsleigh House, Milton Abbot, Devon
Heligan, St Austell, Cornwall
Inverewe Gardens, Gairloch, Scotland
Logan Botanic Gardens, Stranraer, Scotland
Ness Gardens, Chester, Merseyside
Portmeirion, Porthmadog, North Wales
RHS Wisley, Woking, Guildford, Surrey
Rosemoor Garden, Great Torrington, Devon
Royal Botanic Gardens, Edinburgh
Royal Botanic Gardens Kew, Richmond, London
The Hillier Gardens and Arboretum, Romsey, Hampshire
Trebah, Falmouth, Cornwall
University Botanic Gardens, Cambridge

Wakehurst Place, Crawley, Sussex
Younger Botanic Gardens, Benmore, Scotland

Ireland

Annes Grove, Castletownroche, County Cork
Birr Castle, Desmesne, County Offaly
Derreen, Lauragh, County Kerry
Glasnevin Botanic Gardens, Dublin
Kennedy Arboretum, New Ross, County Wexford
Mount Usher, Ashford, County Wicklow
Mucross, Killarney, County Kerry

Netherlands

Bamboekwekerij Fastuosa (see Where to Buy Bamboos)
Bamboekwekerij KIMMEI (see Where to Buy Bamboos)
Bamboepark Schellinkhout, Dorpsweg 125, 1697 Schellinkhout
De Groene Prins, Kwikkels 3 8341 SK Steenwijkerwold
Hortus Botanicus, Leiden
Rotterdam Zoo Botanic Garden
'Hortus Haren', Near Groningen, Rotterdam

Switzerland

Botanic Garden Isole di Brissago, Ticino
Botanic Garden of Basel
Botanic Garden of Geneva
Botanic Garden of the University of Zürich
Centre Horticole de Lullier, near Geneva
Chinese Garden, Zürich
Natural History Museum, Winter Garden, 3005 Bern
Parco Scherrer, Morcote, Ticino
Vallée de Bambou, Geneva

USA

Arnold Arboretum, Jamaica Plain, Massachusetts
Avery Island Jungle Garden, Los Angeles, California
Bamboo Garden, Foothills College, Los Altos Hills, California
Bamboo Germplasm Repository ARS-USDA Research Station, Byron, Georgia
Brooklyn Botanic Garden, New York
Hakone Gardens, Saratoga, California
Huntington Botanic Gardens, San Marino, California
Leu Botanical Gardens, Orlando, Florida
Live Oak Gardens, New Iberia, Los Angeles, California
Longwood Gardens, Kennett Square, Pennsylvania
Los Angeles State and County Arboretum, Arcadia, California
Mercer Arboretum and Botanic Gardens, Humble, Texas
Morris Arboretum, Philadelphia, Pennsylvania
North Carolina State Arboretum, Raleigh, North Carolina
Quail Botanical Gardens, 230 Quail Gardens Drive, Encinitas, California
San Diego Zoo & Wild Animal Park (Central Zoo San Diego), San Diego, California
Strybing Arboretum & Botanical Gardens, San Francisco, California
The San Antonio Zoo, San Antonio, Texas
United States National Arboretum, Washington DC
University of California Botanic Garden, Riverside, California
University of California Botanic Gardens, Berkeley, California
Washington Park Arboretum, Seattle, Washington
Zilker Botanic Gardens, Austin, Texas

APPENDIX 4
WHERE TO BUY BAMBOOS

A few specialist nurseries are given below. Contact your local bamboo society for a more comprehensive list.

France

Bambous du Mandarin, Pont de Siagne, 83440 Montauroux
Bambous de Planbuisson, Rue Montaigne, 24480 Le Buisson
Fleurimont, 5 Chemin Grevileas, 97460 Saint-Paul
Jardin d'Ombre et de Lumière, 9 rue Lafayette, 94210 La Varenne St-Hilaire
Pépinières de la Bambouseraie Prafrance, Générargues 30140 Anduze
Pépinières de Blonzac, 97128 Goyave

Germany

Baumschulen, Nussbaumallee 69, D-6100, Darmstradt
Baumschule, Saarstrasse 3, D-7570, Baden Baden
Sortiments und Versuchsgartnerei, Staudenweg, D-8772, Marktheidenfeld

Britain

Drysdale Garden Exotics, Bowerwood Road, Fordingbridge, Hampshire SP6 1BN
The Japanese Connection, Market Harborough, Leicestershire
Jungle Giants, Burford House, Tenbury Wells, Worcestershire WR15 8HQ
Just Bamboo Ltd, 109 Hayes Lane, Bromley, Kent BR2 9EF
The Rodings Plantery, Plot 3, Anchor Lane, Abbess Roding, Essex CM5 OJW
Scottish Bamboo Nursery, Middlemuir Farm, Craigievar, Alford, Aberdeenshire AB33 8JS
Sunnyside Nurseries, Heath Road, Kenninghall, Norfolk NR16 2DS
Tuckermarsh Gardens, Bere Alston, Yelverton, Devon PL20 7HN

Ireland

Stam's Nursery, The Garden House, Cappoquin, County Waterford

Italy

Firma Baldacchi, Pistoia

Netherlands

Bamboekwekwekerij Fastuosa, Dr Kijlstraweg 46, 9063 JD Molenend
Bamboekwekwekerij KIMMEI, Zandbergstraat 14, 5555 LB Valkenswaard

Switzerland

Alfred Forster, 3207 Golaten, near Kerzers

USA

David C. Andrews, PO Box 358, Oxon Hill, MD 20750-0358
Bamboo Garden Nursery, 1507 S.E. Adler, Portland, OR 97214
Bamboo Gardener, 2609 NW 86th Street, Seattle, WA 98117-3838
Bamboo International, 1902 Boundary Avenue, Ramona, CA 92065
Bamboo Sourcery, 666 Wagnon Road, Sebastopol, CA 95472
Burt Associates Bamboo, PO Box 719, 3 Landmark Road, Westford, MA 01886
Endangered Species, 23280 Stephanie Perris, CA 92570
King's Creek Gardens, 813 Straus Road, Cedar Hill, TX 75104
Louisiana Nursery, 5853 Highway 182, Opelousas, LA 70570
Mike's Garden Center, 5703 Crowley Road, Fort Worth, TX 76134
New England Bamboo Co., 5 Granite St, Rockport, MA 01966
Northern Groves, PO Box 1236, Philomath, Oregon 97370
Our Bamboo Nursery, 30 Myers Road, Summertown, TN 38483-7323
Steve Ray's Bamboo Gardens, 250 Cedar Cliff Road, Springville, AL 35146
Tom Wood, Nurseryman, PO Box 100, Archer, FL 32618
Tradewinds Nursery, 28446 Hunter Creek Loop, Gold Beach, OR 97444
Tripple Brook Farm, 37 Middle Road, Southampton, MA 01073
Upper Bank Nurseries, PO Box 486, Media, PA 19063

APPENDIX 5
BAMBOO SOCIETIES

EUROPE

Activity Co-ordinator
c/o Yolande Younge-Petersen, Dorpsweg 125, NL-1679 KJ Schellinkhout, Netherlands.
tel. (0)229 501970.

EBS (Belgium)
c/o Johan Gielis, Nottebohmstraat 8, B-2018 Antwerpen 1.
tel. +32 3 2364629.

EBS (France)
c/o Martine Bouret, Rue de l'Eglise, 30170 La Rouquette.
tel. +33 4 66 77 00 47.
email. fboure@aol.com

EBS (Germany)
c/o Frau Edeltraud Weber, John-Wesley-Straβe 4, D-63584 Gründau 2 (Rothenbergen).
tel. +49 6051 17451.

EBS (Great Britain)
c/o Martin Brook, 11 Normanhurst, 112 Cherry Hinton Road, Cambridge, CB1 7BJ. *tel.* 07831 111247
email. mb@bamboo-society.org.uk

EBS (Italy)
c/o Mario Brandazzi, Via Dosso di Mattina 19, Ceedera Rubbiano CR.
tel. +39 373 61009

EBS (Netherlands)
c/o Wim Masman, Borculoseweg 2, NL-7261 BJ Ruurlo.
tel. 57 35 14 65.

EBS (Spain)
c/o José Maria Viure, Carretera Cardedeu a Cánoves, km.2 izq, E 08440, Cardedeu.
tel. +34 93 8462001.

EBS (Switzerland)
c/o Magrit Blaser, Strandweg 22, CH Muntelier.

NORTH AMERICA

American Bamboo Society
c/o Susanne Lucas, 9 Bloody Pond Road, Plymouth, MA 02360.
tel. 508 224 7982.
fax. 508 224 4493.
email. slucas0033@aol.com

The American Bamboo Society has many regional chapters. These are listed below. The chapter contacts usually change annually, so it is best to contact the central organization to obtain current information.

Florida Caribbean Chapter
Hawaii Chapter
Louisiana Gulf Coast Chapter
Northeast Chapter
Northern California Chapter
Oregon Bamboo Association
Pacific Northwest Chapter
Southeast Chapter
Southern California Chapter
Texas Chapter
Tierra Seca Chapter

Bamboo Newsletter of Canada
c/o Michael Curtis, 14050 60th Avenue, Surrey BC, V3X 2N3, Canada.

AUSTRALIA AND NEW ZEALAND

Australian Bamboo Network
c/o Peter Bindon, PO Box 174, Fremantle, WA 6160, Australia.
*email.*pbindon@uo3o.aone.net.au

The New Zealand Bamboo Society
c/o Nickie Higgle, PO Box 11, Fordell, Wanganui, New Zealand.

INTERNET CONTACTS

European Bamboo Society
http://www.bodley.ox.ac.uk/users/djh/ebs

American Bamboo Society
http://www.bamboo.org/abs/

Australian Bamboo Network
http://www.ctl.com.au/abn/abn.htm

Mailbox
Bamboo Internet Group

APPENDIX 6
READING ABOUT BAMBOOS

The older books are sometimes available from old book specialists, or modern reprints of most of them can be obtained through the American Bamboo Society Bookstore, 1919 Richcreek Road, Austin, TX 78757. *tel.* 512 452 7146. Contact: Herb Hillery *email.*hrbhillery@aol.com

Bambus Buch, Skalitzerstr. 43, 10997 Berlin, is a bookshop specializing in books on bamboos.

Butt P.P. & others, *Hong Kong Bamboo* (Urban Council, Hong Kong 1985). Descriptions of species grown in this region.

Chao C.S. A *Guide to Bamboos Grown in Britain* (Royal Botanic Garden, Kew 1989).

Dunkelberg K. IL 31 *Bambus Bamboo* (ISBN 3-7828-2031-2). Bamboo as a building material – structural properties and construction techniques.

Farrelly D. *The Book of Bamboo* (Sierra Club Books, San Francisco 1984). Ethnic uses and commercial cultivation.

Freeman-Mitford A.B. *The Bamboo Garden* (Macmillan, London and New York 1896). A classic.

Judziewicz, Clark, Londono, & Stern *American bamboos* (Smithsonian Institute Press 1998). Structure, ecology, use of bamboos.

Lawson A.H. *Bamboos: A Gardener's Guide to their Cultivation in Temperate Climates* (Faber and Faber, London 1968). An essential historic classic.

McClure F.A. *Bamboos of the genus* Phyllostachys *under Cultivation in the United States* (Agricultural handbook No 114, US Department of Agriculture, Washington DC1957). Essential for identifying this genus.

McClure F.A. *The Bamboos – a fresh perspective* (Harvard University, Cambridge, Massachusetts 1966). A historic but relevant book on botanic structure and growth.

Meredith, Ted Jordan, *Bamboo for Gardens* (Timber Press 2001). Comprehensive descriptions of all aspects of bamboo growing in the USA.

Ohrnburger D. & Goerrings J. *The Bamboos of the World.* Preliminary studies of classification with location maps, published by the authors as a series of papers.

Okamura H. *Illustrated Horticultural Bamboo Species in Japan* (HAATO Ltd. Japan 1991). Descriptions of species grown in this region.

Recht C. & Wetterwald M.F. *Bamboos* (Timber Press, Portland 1992). General horticulture.

Satow E. *The Cultivation of Bamboos in Japan* (Transactions of the Asiatic Society of Japan 27 part 3; The Asiatic Society of Japan, Tokyo 1899). An essential historical work, translated and extensively annotated by Satow from the Japanese book *Nikon Chiku-Fu* by Katayama Nawohito.

Stapleton C.M.A. *Bamboos of Bhutan: an illustrated guide* (Royal Botanic Gardens, Kew 1994).

Stapleton C.M.A. *Bamboos of Nepal: an illustrated guide* (Royal Botanic Gardens, Kew 1994). Two very useful books detailing species from these important regions.

Starosta P. & Crouzet Y. *Bamboos* (Benedikt Taschen Verlag GmbH 1998). Beautiful photography with a brief text.

Suzuki O. & Yoshikawa I. *The Bamboo Fences of Japan* (Graphic-sha, Tokyo 1988). Artistic and practical.

Suzuki S. *Index to Japanese Bambusaceae* (Gakken Co Ltd., Tokyo 1978). Descriptions of species grown in this area.

Wang D. & Shen S. *Bamboos of China* (Christopher Helm, London 1987).

Zhu S.L. & others A *Compendium of Chinese Bamboo* (China Forestry Publishing House, Peking 1994). Two complementary descriptive books about Chinese species.

PERIODICALS

Temperate Bamboo Quarterly Adam & Sue Turtle, 30 Myers Road, Summertown, TN 38483-7323, USA. *tel.* 931 964 4151.

APPENDIX 7
BAMBOO SYNONYMS

A few of the more common name changes, confused names or equivalent names. Specific names only. Old names are given on the left with their correct new name on the right.

aff culeou – *Chusquea gigantea*
auricoma – *Pleioblastus viridistriatus*
auricomus – *Pleioblastus viridistriatus*
breviglumis (past classification) – *Chusquea culeou* 'Tenuis' (the dwarf *Chusquea*)
breviglumis (recent classification) – *Chusquea gigantea*
castilloni – *Phyllostachys bambusoides* 'Castillonis'
cernua 'Nebulosa' – *Sasa palmata* 'Nebulosa'
congesta – *Phyllostachys atrovaginata*
falcata – *Drepanostachyum falcatum*
glaucescens – *Bambusa multiplex*
gracilis – *Drepanostachyum falcatum*
gracillima – *Drepanostachyum falcatum*
hayatae – *Sasa veitchii* 'Nana' ('Minor')
heterocycla – *Phyllostachys edulis* 'Heterocycla' (sometimes *P. edulis*)
heterocycla 'Pubescens' – *Phyllostachys edulis*
hookerianum – *Himalayacalamus falconeri* 'Damarapa'
jaunsarensis – *Yushiana anceps*
mitis – *Phyllostachys sulphurea* 'Viridis'
pubescens – *Phyllostachys edulis*
pumila – *Pleioblastus humilus* 'Pumilus'
purpurata – *Phyllostachys heteroclada*
quilioi – *Phyllostachys bambusoides*
'Robert Young' – *Phyllostachys sulphurea* 'Sulphurea'
simonii 'Heterophyllus' – *Pleioblastus simonii* 'Variegatus'
spathacea – *Fargesia murieliae*
spathiflora – *Thamnocalamus spathiflorus*
vagans – *Sasaella ramosa*
variegata – *Pleioblastus fortunei*
variegatus – *Pleioblastus fortunei*
villosa – *Semiarundinaria okuboi*
viridis – *Phyllostachys sulphurea* 'Viridis'
viridula – *Pseudosasa pleioblastoides*

INDEX

Page numbers in *italics* refer to illustrations

ACKNOWLEDGEMENTS

It is impossible to assess the influence of gardening predecessors on my art, and also on this book. I am honoured that the culmination of the aspirations and dreams of the gardeners and land owners of the last century are all around for me to experience. Their works of art can be seen everywhere in Cornwall, and it is the bedrock upon which gardening in this part of the world is built. Bamboos have a very significant place in this history, and can be found in almost every old garden in this region.

The beauty and natural drama of this small part of the world was undoubtedly the catalyst that stimulated my imagination and led to thoughts of a book about the horticulture of bamboos. This small glow was nurtured by my gardening friends who were all enthusiastic about the idea. Particularly supportive were Les Cathery and Anthony and Jane Rogers of Carwinion. Neil Armstrong was a great stimulus and gave the final push to make the idea a reality.

The weeks of routine writing were interspersed with short periods of intense activity when the photographers arrived. Again Anthony and Jane Rogers were most supportive, allowing us unrestricted access to their beautiful garden. Many of the outside photographs are of Carwinion and most of the others were taken in Trebah and Heligan, those other dramatic Cornish gardens. Major Hibbert was also most hospitable, and freely and without hesitation allowed us to photograph the many fine views and plants at Trebah. Similarly, John Nelson and Tim Smit of Heligan were most helpful and co-operative.

From the start I have been very aware of the need to avoid an introverted outlook to this publication and have drawn upon my earlier experiences in less favoured regions. But even this does not equip me to comment constructively about the performance of bamboo in the very cold regions of the world. Correspondence over the years with Max Riedelsheimer, and also recently with Bill Hoag, has been most interesting and instructive, and of great assistance in correcting my shortcomings. In addition Ned Jaquith has advised upon conditions appertaining to the west coast of the North American continent and Susanne Lucas similarly advised on east coast conditions and gave some much appreciated general comments. Jos Kevanah was most helpful with advice and his experiences with bamboos in The Netherlands. Occasionally, experiences were conflicting and served to remind me of the dangers of generalization.

Much time was spent investigating nomenclature and even with the help of the RHS database it was easy to slip into familiar but outdated names. Chris Stapleton and Steve Renvoize from the Royal Botanic Gardens Kew were most helpful in their respective specialities of Himalayan species and the genus *Phyllostachys*. Chris Stapleton was also invaluable in clarifying the finer divisions of classification, and large sections of this subject are taken directly from his interesting letters. Any errors or failures to spot incorrect nomenclature are mine, however.

All of these very good friends have given their assistance without hesitation and my great appreciation is similarly given. I am well aware that without all of their contributions this book would not have been written, or would have been greatly inferior.

The one aspect that has been influential above all others has been a supportive family. This I have been lucky enough to enjoy ceaselessly from the birth of my interest. Janice and Susan, my daughters, have cheerfully and efficiently undertaken the monotonous task of proofreading in spite of their own heavy commitments, and this will always be appreciated. Ann, my wife, was with me on that eventful day at Penjerrick, more years ago than I wish to remember. She has been a positive influence over the years whenever I have indulged myself in my chosen interest. Without her unfailing support, this book would not have been written and it is to her that it is dedicated.